Principles of
Highway Engineering
and Traffic Analysis

Principles of Highway Engineering and Traffic Analysis

Fred L. Mannering
University of Washington

Walter P. Kilareski
The Pennsylvania State University

WILEY

John Wiley & Sons

New York   *Chichester*   *Brisbane*   *Toronto*   *Singapore*

Library of Congress Cataloging in Publication Data:

Mannering, Fred L.
    Principles of highway engineering and traffic analysis/Fred L.
Mannering, Walter P. Kilareski.
      p.   cm.
    Includes bibliographical references.
    ISBN 0-471-63532-4
    1. Highway engineering. 2. Traffic engineering.   I. Kilareski,
Walter P. II. Title.
TE147.M28 1990
625.7—dc20

Printed in the United States of America

10 9 8 7 6 5 4 3 2 1

Preface

Lack of an appropriate textbook is one of the most common complaints voiced by instructors of entry-level transportation engineering courses. Although many fine transportation engineering textbooks are in print, there are few, if any, that are completely suitable for entry-level transportation engineering courses. Virtually all existing textbooks attempt to cover the breadth of the transportation engineering field and thus sacrifice the depth of coverage that most students in entry-level courses expect. Since the entry-level transportation engineering course is frequently the only transportation course required of all civil engineering students, their perception of the transportation field is often one of rambling breadth with minimal coherency. We are convinced that many prospective transportation students are lost to other civil engineering disciplines as a result of the absence of a well-tailored entry-level text.

Principles of Highway Engineering and Traffic Analysis achieves a major departure from previous transportation engineering textbooks by (1) focusing on a single mode, highway transportation, and (2) providing a depth of coverage of individual topics. The focus on the highway mode seems to be a natural one because of dominance of the highway mode in the United States and the fact that undergraduates employed in transportation will probably find themselves working on highway-related problems. Offering a greater depth of coverage is a more risky proposition. In so doing, we have isolated the most fundamental principles of highway engineering and traffic analysis with the intention that, with such principles in hand, individual instructors can easily supplement the text discussion with material that they personally believe deserves additional attention. We believe that such an approach is critical to effective teaching, since it provides students with a well-focused core of material (as represented in the text) while enabling instructors to present material based on their own backgrounds and interests in transportation.

Within the basic philosophical approach described above, we address the common complaint of students that transportation is not as mathematically challenging or rigorous as other civil engineering disciplines. Addressing this issue is a difficult task because transportation engineering presents a dichotomy with regard to mathematical rigor, with relatively simple mathematics used in practice-oriented material and fairly complex mathematics used in research-oriented material. Thus it is common for instructors to either insult students' mathematical knowledge or to exceed it. This book makes a sincere effort to provide that elusive middle ground of mathematical rigor which matches junior and senior engineering students' mathematical abilities. The numerous example problems and end-of-chapter problems are evidence of this effort.

Principles of Highway Engineering and Traffic Analysis was developed from course notes, used as a sole source of reference, for the introductory transportation engineering classes taught at the University of Washington and the Pennsylvania State University. The material presented in these notes (and now contained in this book) is largely responsible for transforming a much maligned course into one of the very best civil engineering introductory courses, as rated by students. We are hopeful that our favorable experiences with this material can now be shared with other instructors.

The book begins with a short introductory chapter that stresses the significance of highway engineering and traffic analysis in our society. This chapter provides the student with a basic overview of the problems facing the field. The chapters that follow are arranged in sequences that focus on highway engineering (Chapters 2, 3, and 4) and traffic analysis (Chapters 5, 6, and 7).

Chapter 2 introduces the basic elements of road vehicle performance. This chapter represents a major departure from the vehicle performance material presented in previous transportation engineering textbooks, in that it is far more involved and detailed. This additional level of detail is justified on two grounds. First, since students own and drive automobiles, they have a basic interest that can be combined with the principles of physics to provide an important link to their freshman and sophomore engineering course work. Traditionally, the absence of such a link has been a criticism of introductory transportation engineering courses. Second, with the incredible on-going advances in vehicle technology, it is more important than ever that civil engineering students understand the principles involved and ultimately the effect of changing vehicle technologies on highway design practice.

Chapter 3 presents current design practices for the geometric alignment of highways. The chapter is highly focused on stopping-sight distance at the expense of factors such as drainage, highway classifications, and so on. However, we believe that there is a great potential for alienating introductory-level students with too many design-oriented details. Hence, we leave the option of supplementing this material to the discretion of the instructor.

Chapter 4 overviews the theory and practice of pavement design. Basic concerns for both rigid and flexible pavement design are discussed and illustrated through examples.

Chapter 5 introduces the basic tools of traffic analysis. Included are the relationships between traffic speed, flow and density, queuing theory, and applications of queuing models to traffic bottlenecks and signalized intersections.

Chapter 6 presents some current standards of traffic analysis as they are used by practicing engineers. Fundamentals are discussed as well as the complexities involved in the analysis problem. The discussion is structured so as to provide the student with an appreciation of the theoretical compromises necessary to implement traffic analysis methods.

Chapter 7, the final chapter, concludes the discussion with an overview of traffic forecasting. In many respects, traffic forecasting typifies transportation engineering as a whole, demonstrating the breadth in terms of factors involved and methodologies used that makes teaching at the introductory level extremely challenging. To resolve this problem of breadth, we concentrate on a single theoretically and mathematically tractable approach to the topic of traffic forecasting and one that, we believe, conveys the underlying principles involved with extreme effectiveness.

Finally, instructors will find the end-of-chapter problems extremely useful in supporting the content and spirit of the book. These problems are precise and challenging, a combination not always found in introductory transportation engineering textbooks.

Fred L. Mannering
Walter P. Kilareski

Contents

Chapter Five
Elements of Traffic Analysis /127

Chapter Six
Applied Traffic Analysis: Basic Elements /167

Chapter One

Introduction to Highway Engineering and Traffic Analysis

1.1 INTRODUCTION

Highway transportation is the critical underpinning upon which the industrial and technological complex of the United States is based. Virtually every aspect of the U.S. economy can be directly tied to highways; from the movement of freight and finished goods, to the transportation of personnel to and from work, shopping centers, and recreational locations, highways are key components to our economic health. It is interesting that the importance of the U.S. highway system is a rather recent development. Prior to the Second World War, much of the nation's freight and passenger intra- and intercity movement was undertaken on rail. Although the popularity of motor transportation continued to grow after the war and at a healthy pace, the construction of the interstate highway system in the 1960s and 1970s truly established highway transportation as the dominant mode of transport in the United States. The interstate system, originally motivated on the basis of national defense needs, provided speed and convenience of highway travel that had not been known before. The interstate system was also an incredible engineering accomplishment and still rates as the largest civil engineering project ever undertaken in the history of mankind.

Given the dominance of the highway transportation mode, civil engineers must strive toward two goals: (1) the provision of a high level of service (i.e., minimize travel times and delay) and (2) the provision of a high level of safety. These two goals are not only sometimes contradictory (e.g., higher speeds minimize travel time but also decrease safety), but must be achieved in the context of ever-changing constraints. Such constraints can be broadly classified as economic (costs of projects) and environmental (impacts of projects on the environment including both noise and air quality impacts). As a further complicating concern, the engineer must also address the likely short- and long-term impacts of highway projects on vehicular traffic, which is an outgrowth of traveler behavior.

In spite of the inherent difficulty encountered in attempting to provide higher and higher levels of highway service and safety, engineers have an ethical responsibility to strive toward these goals. In so doing, they face challenges that are technical and behavioral in nature. In the following sections of this chapter we discuss the important dimensions of these challenges.

1.2 TECHNOLOGICAL CHALLENGES

As with all fields of engineering, technological developments offer the promise of solving complex problems and achieving lofty goals. The quest for technological development or, equivalently, the technological challenges in highway transportation include challenges of infrastructure, vehicle technologies, and traffic operations.

1.2.1 Infrastructure

America made an incredible highway capital investment in the 1960s and early 1970s by constructing the interstate system and undertaking the construction of many other roadways. The economic climate that permitted such an ambitious construction program was indeed unique and is not likely to ever exist again. For example, if the interstate system were to be considered for construction today, it would be readily dismissed as economically infeasible because of the substantial increases in construction costs, land acquisition costs, resource costs, and environmental costs that have occurred over the years. Indeed we were extremely fortunate to have constructed the interstate system at the time when the technology and economics favored such endeavors, since the cost of new highway construction today is prohibitive.

Unfortunately, the costs associated with such an enormous highway system go well beyond initial capital investments. That is, although highways can be viewed as durable long-lasting investments, they still require maintenance and rehabilitation at fairly regular intervals. Most of the interstate system was designed to survive 25 to 30 years before major rehabilitation was necessary. Thus an unfortunate consequence of our rapid highway construction programs of the 1960s and 1970s is that a massive rehabilitation program is required in the 1990s to sustain the highway system. If this rehabilitation is deferred, the system will suffer unacceptable losses in the level of service and safety.

A recent report of the Secretary of Transportation to the United States Congress provided a number of overall trends relating to the condition of the nation's highways [1987]. The report indicated that states have made pavement reconstruction and resurfacing a high priority and, as a consequence, between 1982 and 1985 pavement conditions actually improved at a greater rate than they deteriorated on the interstate system, reversing a decade-long trend of pavement condition decline. It is important to note, however, that the report also indicated that approximately 11 percent of interstate pavement mileage was still considered to be deficient from a condition point of view. Given that the age and use of the interstate system are such that pavement conditions are expected to decline precipitously in years to come, it is doubtful that state and federal funding levels will be sufficiently high to avert a growth in the percentage of deficient pavement mileage.

The challenge confronting engineers in the infrastructure rehabilitation and maintenance area is to develop new techniques and technologies to economically combat our aging infrastructure. Such technological developments are needed in the areas of reconstruction practices and productivity, and materials development and research.

1.2.2 Vehicle Technologies

Until the 1970s, vehicle technologies evolved rather slowly and often in response to vehicle-market trends as opposed to some underlying trend toward technological advancement. In the early 1970s, three factors began a cycle of unparalleled vehicle technological advancement that continues to this day: (1) government regulations concerning air quality and vehicle occupant safety, (2) energy shortages, fuel price increases and governmental regulations mandating vehicle fuel efficiency increases, and (3) intense foreign competition in vehicle markets. The aggregate effect of these factors has resulted in vehicle consumers that are hungry for new technology, and vehicle manufacturers have found themselves scrambling resources and redesigning manufacturing processes to provide this technology.

Although the development of new vehicle technologies is usually in the domain of other engineering disciplines (e.g., mechanical engineering), the influence of such technologies on highway design, and as a consequence civil engineering, is unavoidable. Civil engineers face the important challenge of being able to fully account for new vehicle technologies in the design and rehabilitation of highway systems, thus providing for the highest possible level of service and safety.

1.2.3 Traffic Control

Perhaps the most obvious example of traffic control is the intersection traffic signal. At signalized intersections, the goals of providing the highest possible level of service and safety are brought into sharp focus. Procedures for setting traffic signals (allocating available green times to conflicting traffic movements) has made significant advances in recent years—we now have signals that respond to prevailing traffic flows, groups of signals that are sequenced to allow for a smooth through-flow of traffic, and in some cases, computers controlling entire networks of signals. Nevertheless, a simple drive down virtually any arterial in the country underscores the need for further advancements in traffic control technology.

Aside from traffic signal technology, there are a number of technologies on the horizon with the potential to greatly enhance traffic operations. For example, in-vehicle guidance systems show great promise for relieving urban congestion by providing drivers with accurate information on existing traffic conditions and directing them to the best possible route. The challenge to civil engineers is not only to participate in the development and refinement of all new traffic operation technology but to ensure that it operates effectively in practice. Regrettably, the effective operation of new control technologies has not always been realized in the past.

1.3 BEHAVIORAL CHALLENGES

It is important to recognize that highway traffic, the volume of which directly influences highway level of service and safety, is not a physical phenomenon as are many engineering concerns, but a behavioral phenomenon that is the outgrowth of people's travel-related choices. The combined effect of these travel-related choices has resulted in traffic volumes that have produced crippling congestion in many of our urban areas, so that the future prospects for lack of traffic congestion are not encouraging. The trend for traffic growth (measured in terms of vehicle miles of travel), until the next century, is shown in Fig. 1.1. In this figure, two compound growth rates are used for traffic projections; a 2.15 percent annual rate of increase and a more likely (based on recent traffic trends) 2.85 percent annual rate. Barring any major departure from past and current trends, Fig. 1.1 implies that traffic is likely to grow at an alarming rate. As further evidence of this growth, a recent study measured future congestion prospects [Lindley 1987] by rating congestion among U.S. metropolitan areas using a congestion severity index, defined as the ratio of total vehicle delay (in vehicle-hours, relative to free-flow traffic conditions) to million vehicle-miles of travel. The results of the study (see Table 1.1) show that traffic congestion, already considered unbearable by many, is forecasted to grow substantially in the future. The projected growth in congestion is a serious detriment to the goal of providing higher levels of service.

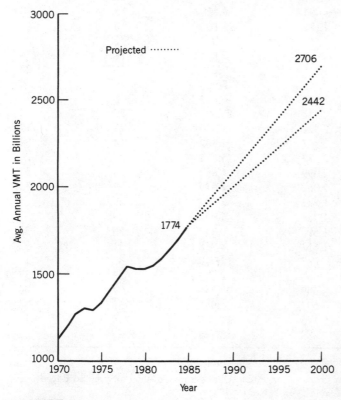

FIGURE 1.1
Vehicles miles of travel in the United States.

TABLE 1.1
Congestion Severity Among U.S. Metropolitan Areas

Urban Area	Index 1984	Index 2005	Ranking 1984	Ranking 2005
Houston	11,112	54,810	1	2
New Orleans	10,576	27,641	2	7
New York	8,168	12,282	3	14
Detroit	7,757	42,394	4	3
San Francisco	7,634	18,734	5	10
Seattle	7,406	27,523	6	8
Los Angeles	6,376	12,139	7	15
Boston	5,538	21,237	8	9
Charlotte	5,263	76,393	9	1
Atlanta	5,034	11,205	10	18
Minneapolis	4,704	9,529	11	21
Dallas	4,630	36,938	12	5
Norfolk	4,505	9,258	13	23
Chicago	4,501	10,700	14	19
Denver	4,454	9,828	15	20
Washington	4,188	15,160	16	11
Hartford	4,111	7,043	17	26
San Antonio	3,938	37,831	18	4
Pittsburgh	3,216	7,243	19	25
San Diego	2,823	5,958	20	28
Cincinnati	2,590	6,223	21	27
Baltimore	2,441	15,037	22	12
Philadelphia	2,421	11,376	23	17
Kansas City	2,347	4,302	24	34
Salt Lake City	2,132	5,811	25	29
Columbus	2,099	4,652	26	33
Cleveland	2,061	4,099	27	35
Sacramento	1,803	8,037	28	24
Milwaukee	1,724	5,653	29	30
Portland	1,696	9,372	30	22
St. Louis	1,612	4,938	31	32
Phoenix	987	12,717	32	13
Providence	660	2,617	33	37
Miami	609	28,549	34	6
Buffalo	577	3,983	35	36
Tampa	575	11,870	36	16
Indianapolis	89	5,148	37	31

Congestion severity index = total delay million vehicle-miles of travel.

Source: J. A. Lindley, "Urban Freeway Congestion: Quantification of the Problem and Effectiveness of Potential Solutions," *ITE Journal*, Vol. 1, No. 1, 1987.

Mitigation of future growth in traffic congestion is a perplexing problem. The current economics of highway construction prohibit any large-scale highway construction program aimed at increasing roadway capacity. Technological advances, as previously discussed, while having the potential to offer some relief, will probably not keep pace with the rapid growth in traffic volumes. This leads us to the more fundamental issue of addressing the behavioral processes and preferences that generate traffic.

1.3.1 Dominance of Single-Occupant Private Vehicles

Of the available urban transportation modes (e.g., bus, commuter train, subway, or private vehicle), private vehicles in general and single-occupant vehicles in particular offer a level of travel mobility that is unequaled. The private vehicle is such an overwhelmingly dominant choice that travelers are willing to pay substantial capital and operating costs, confront high levels of congestion, and struggle with parking-related problems, just to have the flexibility in terms of travel departure times and destination choices uniquely provided by private vehicles. As shown in Table 1.2, the dominance of the private vehicle mode has continued to erode the modal shares of other alternatives in spite of increasing levels of traffic congestion and the increases in vehicle fuel costs brought on by the two energy crises of the 1970s [Eno Foundation 1987]. Another important trend, in terms of congestion mitigation, is the ongoing decline in the already low vehicle occupancy rate, as shown in Table 1.3 [Ibid.]. Figure 1.2 provides further evidence of this trend, as shown by the growth in motor vehicle registration relative to the number of licensed drivers. The trend toward ever lower vehicle occupancy rates is strong evidence of the dominance of the single-occupant private vehicle modal choice.

Dealing with the issue of single-occupant vehicle travel presents the civil engineering profession with a classic dilemma in striving toward its goal of providing ever higher levels of service. On one hand, developing programs that encourage travelers to take public transportation modes (e.g., bus fare incentives

TABLE 1.2
Trends in Modal Shares, 1960–1990 (in percent)

	1960	1970	1980	1990 (est)
Private vehicle	69.5	80.6	85.9	89.2
Public transit	12.6	8.5	6.2	5.1
Walk to work	10.4	7.4	5.6	4.1
Work at home	7.5	3.5	2.3	1.6

Source: Adapted from Eno Foundation for Transportation, Inc., "Commuting in America: A National Report on Commuting Patterns and Trends," Westport, Conn., 1987.

TABLE 1.3
Selected Average Vehicle Occupancy Rates for Commuting (persons per vehicle)

	1970	1980	1990 (est)
U.S. average	1.18	1.15	1.13
All U.S. metro areas	1.17	1.15	1.13
All U.S. nonmetro areas	1.20	1.18	1.15

Source: Adapted from Eno Foundation for Transportation, Inc., "Commuting in America: A National Report on Commuting Patterns and Trends," Westport, Conn., 1987.

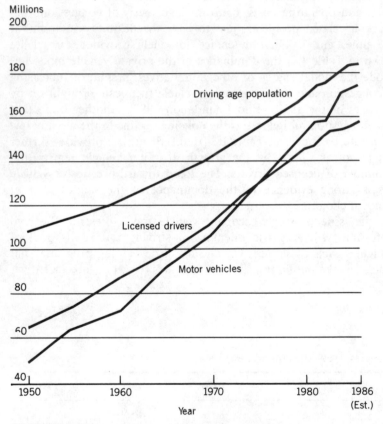

FIGURE 1.2
Trends of driving age population, licensed drivers, and motor vehicle registrations in the United States.

and increases in private vehicle parking fees) or programs to increase vehicle occupancy (e.g., high occupancy vehicle lanes and employer-based ridesharing programs) will tend to relieve traffic congestion on highways and provide highway users with an improved level of service. However, such programs have the adverse effect of directing travelers toward inferior modes that inherently provide lower levels of mobility and, consequently, service. With this dilemma in mind, and an economic environment that does not favor additional highway construction, it is clear that major controversial congestion-related compromises must be reached.

1.3.2 General Demographic Trends

Travelers' commuting patterns are inextricably intertwined with their socioeconomic characteristics, such as age, income, household size, education, and job type, as well as the distribution of residential, commercial, and industrial developments within a region. American metropolitan areas have generally experienced population declines in central cities with rapid growth in suburban areas. In many respects this population shift toward the suburbs was encouraged by the abundance of highway capacity in the late 1950s and 1960s. This suburban shift established the classic commuting pattern of workers living in the suburbs and working in the central city. Conventional wisdom suggested that as overall metropolitan traffic congestion grew, making this commuting pattern much less attractive, commuters would seek to avoid congestion by reverting back to public transport modes or choosing once again to reside in the central cities. However, evidence of another trend has begun to emerge in that employment centers are beginning to develop in the suburbs, thus providing a viable alternative to the congested suburb-to-city commute. Such demographic dynamics present the engineer with an ever moving target that further complicates the problem of providing high levels of service and safety.

In addition to regional shifts in commercial and residential development, there are long-term population/behavioral concerns that must be considered. For example, due to the distribution of births (i.e., the baby boom following the Second World War) and the advances in medical technology that prolong life, the average age of our citizenry continues to increase. Since older people tend to have substantially slower reaction times, engineers must confront the possibility of changing highway design standards and practices to accommodate slower reaction times and the potentially higher variance of reaction times among highway users.

1.3.3 Highway Safety Trends

Since the early 1960s, a tremendous amount of effort and funds have been expended on strategies and projects designed to improve highway safety. Prior to that, the number of accidents, and in particular, the number of highway deaths

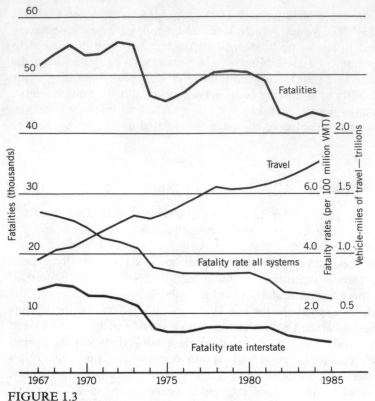

FIGURE 1.3
Motor vehicle fatalities in the United States.

and injuries, were on a tragically increasing trend, with highway fatalities reaching 50,000 plus per year, as shown in Fig. 1.3. Fortunately, recent efforts to improve roadway design and vehicle occupancy protection (e.g., safety belts, padded dashboards, collapsible steering columns, and improved bumper design) have helped to reduce the number of fatalities and the fatality rates (measured by deaths per 100 million vehicle miles traveled, as shown in Fig. 1.3). However, current fatalities still exceed 40,000 per year, which is a steep price to pay for having such a high level of mobility. Many of these deaths can be attributed to behavioral concerns that are beyond the control of the engineering profession, with approximately half of all fatal accidents involving the use of alcohol and another substantial percentage attributable to vehicle occupants not wearing safety belts. Nevertheless, it is the engineer's responsibility to provide the highest level of safety possible within existing economic, physical, and behavioral constraints.

1.4 MEETING THE CHALLENGE

As outlined in the preceding sections, the highway engineering and traffic analysis problem is exceedingly complex and virtually impossible to solve. In spite of its insolvable nature, or perhaps because of it, this problem provides a challenge that is simply unequaled by any other engineering discipline. Meeting this challenge requires the full use of engineers' mathematical and technological expertise as well as the ability to develop imaginative solutions to subsets of the problem. The intent of this book is to provide a basic fundamental foundation that will enable engineering students to better understand and to begin to fully appreciate the many challenging aspects of the highway engineering and traffic analysis problem, with the expectation that they may someday use their talents to address this problem.

REFERENCES

1. "The Status of the Nation's Highways, Report of the Secretary of Transportation to the United States Congress," U.S. Government Printing Office, Washington, D.C., 1987.

2. J. A. Lindley, "Urban Freeway Congestion: Quantification of the Problem and Effectiveness of Potential Solutions," *ITE Journal*, Vol. 57, No. 1, 1987.

3. Eno Foundation for Transportation, Inc.," Commuting in America: A National Report on Commuting Patterns and Trends," Westport, Conn., 1987.

Chapter Two

Road Vehicle Performance

2.1 INTRODUCTION

A basic understanding of the performance of vehicles forms a valuable foundation for the comprehension and analysis of the many design standards used in highway and traffic analysis. For example, in highway design, the determination of the length of freeway and arterial acceleration lanes, maximum highway grades, stopping sight distances, passing sight distances, and numerous accident prevention devices all rely on a basic knowledge of vehicle performance. Similarly, vehicle performance is a major consideration in the selection and design of traffic control devices, the determination of speed limits, and the timing and coordination of traffic signals.

The study of road vehicle performance serves two important functions. First, it provides an understanding of the highway design and operational control compromises necessary to accommodate the wide variety of vehicles (from high-speed sports cars to heavily ladened trucks) typically encountered. Second, it forms a basis from which the impact of advancing vehicle technologies on existing highway design standards can be assessed. This second function is particularly important in light of the ongoing unprecedented advances in vehicle technology. Such advances will necessitate more frequent updating of design standards as well as more knowledgeable engineers.

The objective of this chapter is to provide an introduction to some of the basic elements of road vehicle performance. In so doing, attention will be given only to the straight-line performance of vehicles (acceleration, deceleration, top speed, and the ability to ascend grades). The cornering performance of vehicles will be overviewed in Chapter 3, but detailed presentations of this material are better suited to more specialized sources [Campbell 1978; Wong 1978; Brewer and Rice 1983].

2.2 TRACTIVE EFFORT AND RESISTANCE

Tractive effort and resistance are the two primary opposing forces that determine the straight-line performance of road vehicles. Tractive effort is simply the force available, at the roadway surface, to perform work and is expressed in pounds. Resistance (also expressed in pounds) is defined as the force impeding vehicle motion. The three major sources of vehicle resistance can be identified as: (1) aerodynamic resistance, (2) rolling resistance (which originates from the roadway surface/tire interface), and (3) grade or gravitational resistance. To illustrate these forces, consider the vehicle force diagram shown in Fig. 2.1. In this figure, R_a is the vehicle aerodynamic resistance (in pounds), R_{rf} is the rolling resistance of front tires (in pounds), R_{rr} is the rolling resistance of rear tires (in pounds), F_f is the tractive effort of front tires (in pounds), F_r is the tractive effort of rear tires (in pounds), W is the total vehicle weight (in pounds), θ_g is the angle of grade (in degrees), m is the vehicle mass (slugs), and a is the rate of acceleration (ft/sec^2).

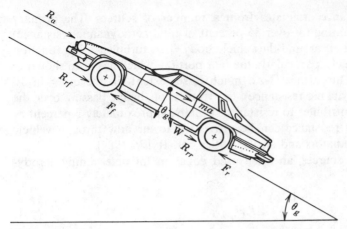

FIGURE 2.1
Forces acting on a road vehicle.

Summing the forces along the vehicle's longitudinal axis provides the basic equation of vehicle motion

$$F_f + F_r = ma + R_a + R_{rf} + R_{rr} + R_g \tag{2.1}$$

where R_g is the grade resistance and is equal to $W \sin \theta_g$. For exposition purposes it is convenient to let F be the sum of the available tractive effort delivered by the front and rear tires ($F_f + F_r$) and similarly to let R be the sum of rolling resistance ($R_f + R_r$). This notation allows Eq. 2.1 to be written as

$$F = ma + R_a + R_r + R_g \tag{2.2}$$

Sections 2.3 to 2.8 of this chapter will present a thorough discussion of the components and implications of Eq. 2.2.

2.3 AERODYNAMIC RESISTANCE

Aerodynamic resistance is a fundamental force that can have significant impacts on vehicle performance. At high speeds, where this component of resistance can become overwhelming, proper vehicle aerodynamic design is essential. Attention to aerodynamically efficient design has long been the rule in race and sports car design and, more recently, concerns over fuel efficiency have resulted in more efficient aerodynamic designs in common passenger cars.

Aerodynamic resistance originates from a number of sources. The primary source (typically accounting for over 85 percent of total aerodynamic resistance) is the turbulent flow of air around the vehicle body. This turbulence is a function of the shape of the vehicle, particularly the rear portion, which has been shown to be the major area of turbulence. To a much lesser extent (in the order of 10 percent of total aerodynamic resistance), the friction of the air passing over the body of the vehicle contributes to resistance. Finally, approximately 3 percent of the total aerodynamic resistance can be attributed to air flow through vehicle components such as radiators and air vents [Scribor–Rylski 1984].

Based on the these sources, an often used equation for determining aerodynamic resistance is

$$R_a = \frac{\rho}{2} C_D A_f V^2 \tag{2.3}$$

Where ρ is the air density in slugs/ft^3, C_D is the coefficient of drag and is unitless, A_f is the vehicle frontal area (projected area of the vehicle in the direction of travel) in ft^2, and V is the speed of the vehicle in feet per second (fps). To be truly accurate, for aerodynamic resistance computations, V is actually the speed of the vehicle relative to the prevailing wind speed. To simplify the exposition of concepts soon to be presented, the wind speed is assumed to be equal to zero for all problems and derivations in this book.

Air density is a function of both elevation and temperature, as indicated in Table 2.1. Equation 2.3 indicates that as the air becomes more dense, total aerodynamic resistance increases. The coefficient of drag, C_D, is a term that implicitly accounts for all three of the air resistance sources discussed above. The coefficient of drag is measured from empirical data either from wind tunnel experiments or actual field tests in which a vehicle is allowed to decelerate from a known speed with other sources of resistance (rolling and grade resistances) accounted for. Table 2.2 gives some approximation of the range of drag coefficients for different types of road vehicles. Table 2.3 presents drag coefficients for specific automobiles covering the period from the late 1960s to the late 1980s. The general trend toward lower drag coefficients over this time period reflects the

TABLE 2.1
Typical Values of Air Density Under Specified Atmospheric Conditions

Altitude (ft)	Temperature (°F)	Pressure (psia)	Air Density (slugs / ft^3)
0	59.0	14.7	0.002378
5,000	41.2	12.2	0.002045
10,000	23.4	10.1	0.001755

TABLE 2.2
Ranges of Drag Coefficients for
Typical Road Vehicles

Vehicle Type	Drag Coefficient (C_D)
Automobile	0.25–0.55
Bus	0.5–0.7
Tractor-trailer	0.6–1.3
Motorcycle with rider	0.6–1.8

enormous concern of an automobile industry struggling with increasing energy costs and manufacturer competition.

The impressive drag coefficient of the Ford Probe V provides some evidence on the practical limit on drag coefficient reduction. Figure 2.2 illustrates the effect of automobile operating conditions on drag coefficients [Janssen and Hucho 1973]. As indicated, even minor factors, such as the opening of windows, can have a significant impact on the drag coefficient and thus total aerodynamic resistance. Projected frontal areas, which typically range from 11 ft^2 to 25 ft^2 for passenger cars, are also a major factor in determining aerodynamic resistance.

TABLE 2.3
Drag Coefficients of Select Automobiles

Vehicle	Drag Coefficient (C_D)
1968 Chevrolet Corvette	0.50
1968 Volkswagen Beetle	0.46
1968 Mercedes 300SE	0.39
1978 Triumph TR7	0.40
1978 Jaguar XJS	0.36
1987 Honda Civic DX	0.35
1987 Acura Integra	0.34
1987 Porsche 944 Turbo	0.33
1987 Ford Taurus	0.32
1987 Mazda RX7	0.31
Ford Probe V (experimental four-passenger sedan)	0.137

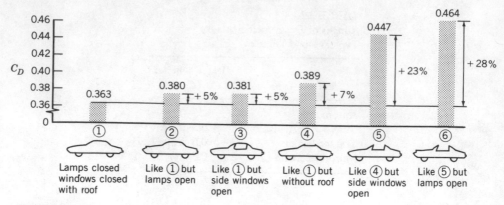

FIGURE 2.2
Effect of operational factors on the drag coefficient of an automobile. (Reproduced by permission from L. J. Janssen and W. H. Hucho, "The Effect of Various Parameters on the Aerodynamic Drag of Passenger Cars" *Advances in Road Vehicle Aerodynamics 1973*, BMRA Fluid Engineering, Cranfield, England.)

Since aerodynamic resistance is proportional to the square of the vehicle speed, it is clear that such resistance will increase rapidly at higher speeds. The magnitude of this increase can be underscored by considering an expression for the power (P_{R_a}) required to overcome aerodynamic resistance. Since power is the product of force and velocity, the multiplication of Eq. 2.3 gives

$$P_{R_a} = \frac{\rho}{2} C_D A_f V^3 \tag{2.4}$$

or, since 1 horsepower is equal to 550 ft-lb per second,

$$\mathrm{hp}_{R_a} = \frac{\rho C_D A_f V^3}{1100} \tag{2.5}$$

Thus the amount of power required to overcome aerodynamic resistance increases with the cube of velocity indicating, for example, that eight times as much power is required to overcome aerodynamic resistance if the vehicle speed is doubled.

2.4 ROLLING RESISTANCE

Rolling resistance, as defined herein, refers to the resistance generated from a vehicle's pneumatic tires. The primary source of such resistance is the natural deformation of the tire body as it passes over the roadway surface. The work needed to overcome this deformation accounts for approximately 90 percent of the total rolling resistance. Depending on the vehicle weight and the characteristics of the roadway surface, the penetration of the tire into the surface and the corresponding surface compression can also be a significant source of rolling resistance. However, for typical vehicle weights and pavement types, penetration/compression only constitutes around 4 percent of the total rolling resistance. Finally, frictional motion due to the slippage of the tire on the roadway surface and, to a lesser extent, air circulation around the tire and wheel (i.e., the "fanning effect") are sources accounting for roughly 6 percent of the total rolling resistance [Taborek 1957].

Considering these sources, three factors affecting rolling resistance are worthy of note. First, the rigidity of the tire and roadway surface will influence the degree of tire penetration, surface compression, and tire deformation. Hard, smooth, and dry roadway surfaces provide the lowest rolling resistance. Second, tire conditions, including inflation pressure and temperature, can also have a substantial impact on the rolling resistance. High tire inflation on hard paved surfaces decreases rolling resistance, as a result of reduced friction, but increases rolling resistance on soft unpaved surfaces due to additional surface penetration. Also, higher tire temperatures make the tire body more flexible and thus less resistance is encountered during tire deformation. The third and final factor, vehicle operating speed, affects tire deformation, with increasing speed producing additional tire flexing and vibration and consequently a higher rolling resistance.

With the diversity of factors affecting rolling resistance considered, studies have shown that rolling resistance can be approximated as the product of a friction term (coefficient of rolling resistance) and the weight of the vehicle acting normal to the roadway surface. The coefficient of rolling resistance (f_r), for road vehicles operating on typical paved surfaces, can be estimated from [Taborek 1957]

$$f_r = 0.01\left(1 + \frac{V}{147}\right) \tag{2.6}$$

where V is the speed in feet per second (fps). By inspection of Fig. 2.1, the rolling resistance (pounds) will simply be the coefficient of rolling resistance multiplied by $W \cos \theta_g$, the vehicle weight acting normal to the roadway surface. For most highway applications, θ_g is quite small, so it can be safely assumed that $\cos \theta_g = 1$, giving the equation for rolling resistance as

$$R_r = f_r W \tag{2.7}$$

From this, the amount of horsepower required to overcome rolling resistance is

$$\mathrm{hp}_{R_r} = \frac{f_r W V}{550} \tag{2.8}$$

Example 2.1

A 2500-lb car is driven at sea level $\rho = 0.002378$ slugs/ft^3 on a concrete surface and has $C_D = 0.38$ and 20 ft^2 of frontal area. It is known that at maximum speed, 50 hp are being expended to overcome rolling and aerodynamic resistance. Determine the car's maximum speed.

Solution

It is known that at maximum speed (V_m),

$$\text{available horsepower} = R_a V_m + R_r V_m$$

or

$$hp = \frac{\left[(\rho/2) C_D A_f V^3 + Wf_r V\right]}{550}$$

Substituting, we obtain

$$27,500 = \frac{0.002378}{2}(0.38)(20) V^3 + 2500[0.01 + 0.000068V]V$$

or

$$27,500 = 0.00904V^3 + 0.17V^2 + 25V$$

Solving for V_m by trial and error gives

$$\underline{\underline{V = 133 \text{ fps},}} \quad \text{or} \quad \underline{\underline{90 \text{ mph}}}$$

2.5 GRADE RESISTANCE

Grade resistance is simply the gravitational force resisting vehicle motion. As suggested in Fig. 2.1, the expression for grade resistance (R_g) is

$$R_g = W \sin \theta_g \qquad (2.9)$$

Again, as was the case in the development of the rolling resistance formula (Eq. 2.7), highway grades are usually small, so that $\sin \theta_g \approx \tan \theta_g$. Rewriting Eq. 2.9, we obtain

$$R_g \approx W \tan \theta_g = WG \qquad (2.10)$$

where G is the grade defined as the vertical rise per some specified horizontal distance (i.e., opposite of the force triangle, Fig. 2.1, divided by the adjacent). Grades are generally specified in percent for ease of understanding. Thus a roadway that rises 5 ft vertically per 100 horizontal ft ($G = 0.05$ and $\theta_g = 2.86°$) is said to have a 5 percent grade.

Example 2.2

A 2000-lb car is traveling at an elevation of 5000 ft ($\rho = 0.002045$ slugs/ft^3) on a concrete surface. If the car is traveling at 70 mph and has $C_D = 0.4$ and $A_f = 20$ ft^2 and an available tractive effort of 255 lb, what is the maximum grade that this car could ascend and maintain the 70-mph speed?

Solution

At maximum speed, tractive effort will be exactly equal to the summation of resistances, with none remaining for vehicle acceleration (i.e., $ma = 0$). Thus Eq. 2.2 can be written as

$$F = R_a + R_r + R_g$$

for grade resistance,

$$R_g = WG$$
$$= 2000G$$

for aerodynamic resistance (using Eq. 2.3),

$$R_a = \frac{\rho}{2} C_D A_f V^2$$
$$= \frac{0.002045}{2}(0.4)(20)(70 \times 1.47)^2$$
$$= 86.61 \text{ lb}$$

and for rolling resistance (using Eq. 2.7),

$$R_r = f_r W$$
$$= \left[0.01\left(1 + \frac{70 \times 1.47}{147}\right)\right] \times 2000$$
$$= 34 \text{ lb}$$

Therefore,

$$F = 255$$
$$= 86.61 + 34 + 2000G$$
$$G = \underline{\underline{0.067}} \quad \text{or a 6.7 percent grade}$$

2.6 AVAILABLE TRACTIVE EFFORT

With the resistance terms in the basic equation of motion (Eq. 2.2) discussed, attention is now directed toward available tractive effort (F) as used in Example 2.2. The tractive effort available to overcome resistance will be determined by either the force generated by the engine or by some maximum value that will be a function of the vehicle weight distribution and the characteristics of the roadway surface/tire interface. The basic concepts underlying these two determinants of tractive effort will be discussed in the following subsections.

2.6.1 Maximum Tractive Effort

Irrespective of the amount of horsepower a vehicle's engine can develop, there is a point beyond which additional power will merely result in the spinning of tires and no additional tractive effort will be generated to overcome resistance. To understand the determinants of this point of maximum tractive effort, a force and moment generating distance diagram is presented in Fig. 2.3, where L is the wheelbase, h is the height of the center of gravity above the roadway surface, l_1 is the distance of the center of gravity from the front axle (l_2 is the distance from the rear axle), W_f is the weight of the vehicle on the front axle (W_r is the weight on the rear axle), and the other terms are as defined for Fig. 2.1.

To determine the maximum tractive effort that the roadway surface/tire contact can support, it is necessary to examine the normal loads on the axles. The normal load on the rear axle (W_r) is given by summing the moments about point A (in Fig. 2.3),

$$W_r = \frac{R_a h + W l_1 \cos \theta_g + mah \pm Wh \sin \theta_g}{L} \tag{2.11}$$

In this equation the grade moment ($Wh \sin \theta_g$) is positive for vehicles on an upward slope and negative on a downward slope. Rearranging terms, assuming $\cos \theta_g = 1$, and substituting into Eq. 2.2 gives

$$W_r = \frac{l_1}{L} W + \frac{h}{L}(F - R_r) \tag{2.12}$$

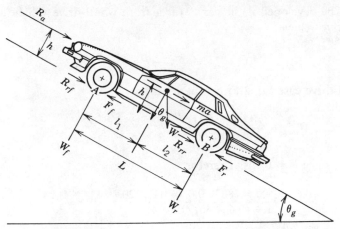

FIGURE 2.3
Vehicle forces and moment generating distances.

From basic physics, the roadway surface/tire maximum tractive effort will be the normal force multiplied by the coefficient of road adhesion,

$$F_{max} = \mu W_r \qquad (2.13)$$

Thus, for a vehicle with rear-wheel drive (substituting Eq. 2.12 into Eq. 2.13), we find that

$$F_{max} = \mu \left[\frac{l_1}{L} W + \frac{h}{L} (F_{max} - R_r) \right]$$

or

$$F_{max} = \frac{\mu W (l_1 - f_r h)/L}{1 - \mu h/L} \qquad (2.14)$$

Similarly, by summing moments about point B, it can be shown that for a front-wheel-drive vehicle

$$F_{max} = \frac{\mu W (l_2 + f_r h)/L}{1 + \mu h/L} \qquad (2.15)$$

Example 2.3

A 2500-lb car is designed with a 120-in. wheel base. The center of gravity is located 40 in. above the pavement and 40 in. behind the front axle. If the coefficient of road adhesion is 0.6 and the car is at rest, what is the maximum tractive effort that can be developed if the car is (a) a front-wheel-drive vehicle and (b) a rear-wheel-drive vehicle?

Solution

(a) For the front-wheel-drive case Eq. 2.15 can be used,

$$F_{max} = \frac{\mu W(l_2 + f_r h)/L}{1 + \mu h/L}$$

From Eq. 2.6 $f_r = 0.01$, since $V = 0$ fps. Therefore,

$$F_{max} = \frac{[0.6 \times 2500 \times (80 + 0.01(40))]/120}{1 + (0.6 \times 40)/120}$$

$$= 837.5 \text{ lb}$$

(b) For the rear-wheel-drive case, Eq. 2.14 can be used:

$$F_{max} = \frac{[0.6 \times 2500 \times (40 - 0.01(40))]/120}{1 - (0.6 \times 40)/120}$$

$$= 618.75 \text{ lb}$$

2.6.2 Engine-Generated Tractive Effort

The amount of tractive effort that can be generated by an engine is a function of a wide variety of engine design factors, including the shape of the combustion chamber, the quantity of air drawn into the combustion chamber during the induction phase, the type of fuel used, and the fuel intake design. Although a complete description of engine design is beyond the scope of this book, an understanding of how engine output is measured and used is important to the study of vehicle performance. The two most commonly used measures of engine output are torque and horsepower. Torque is the work generated by the engine (the twisting moment) and is expressed in foot-pounds (ft-lb). Horsepower is the rate of engine work and is related to torque by the following equation

$$\text{hp}\left(550\frac{\text{ft-lb}}{\text{sec}}\right) = \frac{\text{torque (ft-lb)}}{550} \times \frac{\text{engine rpm}}{60\,(\text{sec/min})} \times 2\pi \qquad (2.16)$$

where engine rpm refers to the number of engine crankshaft revolutions per minute. Figure 2.4 presents a torque–horsepower diagram for a typical gasoline-powered engine.

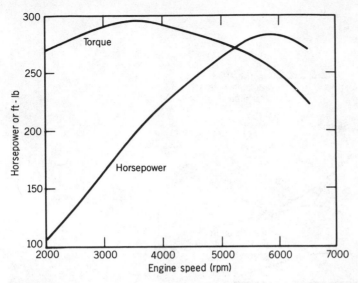

FIGURE 2.4
Typical torque–horsepower curves for a gasoline-powered automobile engine.

There is a direct relationship between engine-generted torque and the tractive effort ultimately delivered to the driving wheels. Unfortunately, the tractive effort needed for accceptable vehicle performance (i.e., to provide adequate acceleration characteristics) is greatest at lower vehicle speeds and, since maximum engine torque is developed at fairly high engine speeds, the use of gasoline engines requires some form of gear reduction as illustrated in Fig. 2.5. This gear reduction provides the mechanical advantage necessary for acceptable vehicle acceleration.

With gear reductions, there are two factors that will determine the amount of tractive effort reaching the driving wheels. First is the efficiency of the gear reduction devices (transmission and differential). Typically, 10 to 25 percent of the tractive effort generated by the engine is lost in the gear reduction devices, which corresponds to a mechanical efficiency (η_t) of 0.75 to 0.90 [Taborek 1957]. Second is the overall gear reduction ratio (ε_0) that plays a key role in the determination of tractive effort. By definition, a gear reduction ratio refers to the relationship between the engine revolutions and road wheel revolutions. For example, a gear reduction ratio of four ($\varepsilon_0 = 4$) means that the engine turns four revolutions per every one turn of the road wheel.

With these terms defined, the engine-generated tractive effort reaching the driving wheels (F_e) is given as

$$F_e = \frac{M_e \varepsilon_0 \eta_t}{r} \tag{2.17}$$

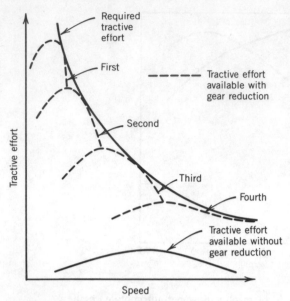

FIGURE 2.5
Tractive effort requirements and tractive effort generated by a typical gasoline-powered vehicle.

where M_e is the torque output of the engine in ft-lb and r is the radius of the wheel in feet. It follows that the relationship between vehicle speed and engine speed is

$$V = \frac{2\pi r n_e}{\varepsilon_0}(1 - i) \tag{2.18}$$

where V is the vehicle speed (fps), n_e is the engine speed in revolutions per second, and i is the natural slippage of the vehicle's running gear generally taken as 2 percent ($i = 0.02$) to 5 percent ($i = 0.05$) for passenger cars [Wong 1978].

In conclusion, the available tractive effort (F in Eq. 2.2) at any given speed will always be the lesser of the maximum tractive effort ($F_{\max}$) and the engine-generated tractive effort (F_e).

2.7 VEHICLE ACCELERATION

Available tractive effort (F), as defined in the previous section, can be used to determine a number of vehicle performance characteristics including vehicle acceleration and top speed. For determining vehicle acceleration characteristics,

Eq. 2.2 can be applied with an additional term added to account for the inertia of the vehicle's rotating parts that must be overcome during acceleration. This term is referred to as the mass factor (γ_m) and is introduced in Eq. 2.2 as

$$F - \Sigma R = \gamma_m ma \qquad (2.19)$$

The mass factor can be approximated by

$$\gamma_m = 1.04 + 0.0025\varepsilon_0^2 \qquad (2.20)$$

Two measures of vehicle acceleration are worthy of note, the time to accelerate and the distance to accelerate. For both, the force (or tractive effort) available to accelerate is $F_{net} = F - \Sigma R$. The basic relationship between the force available to accelerate, F_{net}, the available tractive effort, F (the lesser of F_{max} and F_e), and the summation of vehicle resistances is illustrated in Fig. 2.6. In this figure, F_{net} will be the distance between the lesser of the F_{max} and the F_e curves and the total resistance curve. So, referring to Fig. 2.6, at speed V', F_{net} will be $F_{max} - \Sigma R$ and at V'', F_{net} will be $F_e - \Sigma R$. It follows that when $F_{net} = 0$, the vehicle cannot

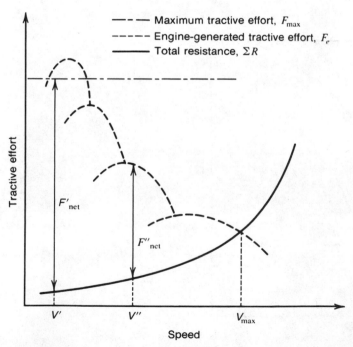

FIGURE 2.6
Relationship among the forces available to accelerate, available tractive effort, and total vehicle resistance.

accelerate and is at its maximum speed for specified conditions (i.e., grade, air density, engine torque, and so on). Such was the case for the vehicle described in Example 2.2. When F_{net} is greater than zero (i.e., the vehicle is traveling less than the maximum speed), Eq. 2.19 can be written in differential form

$$F_{net} = \gamma_m m \frac{dV}{dt} \qquad \text{or} \qquad dt = \frac{\gamma_m m \, dV}{F_{net}}$$

and since F_{net} is itself a function of speed ($F_{net} = f(V)$), integration gives the time to accelerate as

$$t = \gamma_m m \int_{V_1}^{V_2} \frac{dV}{f(V)} \tag{2.21}$$

where V_1 is the initial vehicle speed and V_2 is the final vehicle speed. Similarly, it can be shown that the distance to accelerate is

$$D_a = \gamma_m m \int_{V_1}^{V_2} \frac{V \, dV}{f(V)} \tag{2.22}$$

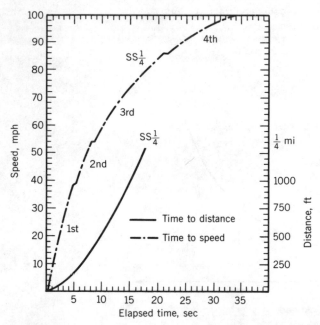

FIGURE 2.7
Example of time-to-speed and time-to-distance curves. (Reproduced by permission from *Road & Track*, MGC Road Test, May 1969.)

Since the integrals expressed in Eqs. 2.21 and 2.22 do not lend themselves to closed-form solutions, numerical integration is necessary. Such integration is straightforward but requires a level of computer programming that is beyond the scope of this book. However, a typical solution to such integration would produce time-to-speed and time-to-distance results similar to those illustrated in Fig. 2.7 [Road & Track 1969].

Example 2.4

On a cold snowy night, a car is going 10 mph ($C_D = 0.3$, $A_f = 20$ ft, $W = 3000$ lb, $\rho = 0.002045$ slugs/ft^3). Its engine is producing 110 ft-lb of torque; the gear reduction ratio is 4.5 to 1; the wheel radius is 14 in.; and the gear reduction efficiency is 80 percent. The road has a concrete surface and the coefficient of road adhesion is 0.2. The driver decides to accelerate to avoid an accident. If the car has a wheelbase of 120 in. and a center of gravity 40 in. above the roadway surface and 50 in behind the front axle, what would the acceleration rate be (a) if the car was front-wheel drive, and (b) if the car were rear-wheel drive?

Solution

It is first necessary to compute the tractive effort generated by the engine, which will be the same for front- and rear-wheel-drive cars.

From Eq. 2.3, we find that

$$R_a = \frac{0.002045}{2}(0.30)(20)(10 \times 1.47)^2$$

$$= 1.33 \text{ lb}$$

from Eq. 2.7,

$$R_r = f_r W$$

$$= 0.01\left(1 + \frac{10 \times 1.47}{147}\right) \times 3000$$

$$= 0.011 \times 3000$$

$$= 33.0 \text{ lb}$$

from Eq. 2.17,

$$F_e = \frac{110(4.5)(0.8)}{\frac{14}{12}}$$

$$= 339.4 \text{ lb}$$

and from Eq. 2.20,

$$\gamma_m = 1.04 + 0.0025(4.5)^2$$
$$= 1.091$$

Therefore, from Eq. 2.19, the engine-generated acceleration is

$$a = \frac{F - \Sigma R}{\gamma_m m}$$

$$= \frac{339.4 - 34.33}{1.091 \times 3000/32.2}$$

$$= 3.0 \text{ ft/sec}^2$$

Whether or not this acceleration rate can be realized depends on the maximum tractive effort that the roadway surface/tire interface can support (as discussed in Section 2.6.1). For the case of the front-wheel-drive car, Eq. 2.15 can be applied

$$F_{max} = \frac{0.2(3000)(70 + 0.011(40))/120}{1 + 0.2(40)/120}$$

$$= 330.1 \text{ lb}$$

which gives an acceleration rate of

$$a = \frac{330.1 - 34.33}{1.091 \times 3000/32.2}$$

$$= 2.91 \text{ ft/sec}^2$$

Since this value is less than the acceleration that the engine is capable of generating, the answer to Part (a) will be the maximum tractive effort acceleration of 2.91 ft/sec^2, as governed by the pavement/tire interface.

For the real-wheel-drive case, Eq. 2.14 is used,

$$F_{max} = \frac{0.2(3000)(50 - 0.11(40))/120}{1 - 0.2(40)/120}$$

$$= 265.5 \text{ lb}$$

$$a = \frac{265.5 - 34.33}{1.091 \times 3000/32.2}$$

$$= 2.27 \text{ ft/sec}^2$$

Since this value is again less than the engine-generated acceleration, the acceleration rate for the rear-wheel-drive car [answer to Part (b)] will be 2.27 ft/sec^2.

2.8 FUEL ECONOMY

Recent concerns over the global availability of fuel have prompted vehicle manufacturers to focus on vehicle design improvements that result in improved fuel efficiency. Based on the information provided in the previous sections of this chapter, some significant factors affecting a vehicle's fuel efficiency can be identified. The most obvious relates to the generation of tractive effort. Engine designs that increase the quantity of air entering the combustion chamber, decrease internal engine friction, and improve fuel delivery and ignition all lead to improved fuel efficiency. Improvements to mechanical components (gear reduction devices) including decreasing slippage and improving mechanical efficiency, can also lead to increased efficiency [Hillard and Springer 1984; Society of Automotive Engineers 1982; Institution of Mechanical Engineers 1983].

In terms of resistance-reducing options, decreasing the overall weight will lower grade and rolling resistances, thus lessening fuel consumption (all other factors held constant). Similarly, aerodynamic improvements such as lower drag coefficients (C_D) reduced frontal areas (A_f) can result in significant fuel savings. Finally, improved tire designs with lower rolling resistances and operating vehicles at lower speeds can also lead to fuel savings.

2.9 PRINCIPLES OF BRAKING

Perhaps the most important highway and traffic related aspect of vehicle performance is braking. The braking behavior of vehicles is critical in the determination of adequate stopping sight distance, roadway surface design, and accident avoidance systems. Moreover, ongoing advances in braking technology make it essential that transportation engineers have a basic comprehension of the principles involved.

2.9.1 Braking Forces

To begin the discussion of braking principles, consider the force and moment generating distance diagram in Fig. 2.8, where F_{bf} and F_{br} are the front and rear braking forces, respectively, and other terms are as previously defined. During vehicle braking there is a load transfer from the rear to the front axle. To illustrate this, expressions for the normal loads on the front and rear axles can be written by summing moments about roadway surface/tire contact points A and B (as was done in deriving Eq. 2.12, with $\cos \theta_g$ assumed to be equal to one):

$$W_f = \frac{1}{L}[Wl_2 + h(ma - R_a \pm W \sin \theta_g)] \qquad (2.23)$$

and

$$W_r = \frac{1}{L}[Wl_1 - h(ma - R_a \pm W \sin \theta_g)] \qquad (2.24)$$

where, in this case, the contribution of grade resistance ($W \sin \theta_g$) is negative for uphill grades and positive for downhill grades.

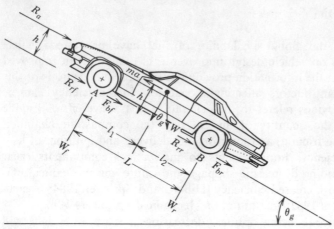

FIGURE 2.8
Forces acting on a vehicle during braking with driveline resistance ignored.

Also, from the summation of forces along the vehicle's longitudinal axis

$$F_b + f_r W = ma - R_a \pm W \sin \theta_g \qquad (2.25)$$

where $F_b = F_{bf} + F_{br}$. Substituting Eq. 2.25 into Eqs. 2.23 and 2.24 gives

$$W_f = \frac{1}{L}[Wl_2 + h(F_b + f_r W)] \qquad (2.26)$$

and

$$W_r = \frac{1}{L}[Wl_1 - h(F_b + f_r W)] \qquad (2.27)$$

Since the maximum vehicle braking force ($F_{b\,max}$) is equal to the coefficient of road adhesion, μ, multiplied by the weights normal to the roadway surface, then

$$F_{bf\,max} = \mu W_f$$
$$= \frac{\mu W[l_2 + h(\mu + f_r)]}{L} \qquad (2.28)$$

and

$$F_{br\,max} = \mu W_r$$
$$= \frac{\mu W[l_1 - h(\mu + f_r)]}{L} \qquad (2.29)$$

To develop maximum braking forces, the tires should be at the point of an impending slide. If the tires begin to slide, a significant reduction in the coefficient of road adhesion will occur. An indication of the extent of the reduction in road adhesion as a result of tire slide, under various pavement types and weather conditions, is presented in Table 2.4 [Shadle, Emery, and Brewer

TABLE 2.4
Typical Values of Coefficients of
Road Adhesion for Vehicle Braking

| | Coefficient of Road Adhesion | |
Pavement	Maximum	Slide
Good, dry	1.0	0.80
Good, wet	0.90	0.60
Poor, dry	0.80	0.55
Poor, wet	0.60	0.30
Packed snow or ice	0.25	0.10

Source: S. G. Shadle, L. H. Emery, and H. K. Brewer, "Vehicle Braking, Stability and Control," *SAE Transactions*, Vol. 92, paper 830562, 1983.

1983]. It is clear from this table that braking forces will decline dramatically when the wheels are locked (resulting in tire slide) during panic stops.

2.9.2 Brake Force Proportioning and Efficiency

In the design of vehicle braking systems, it is necessary to distribute braking forces between the vehicle's front and rear brakes. This is typically achieved by the allocation of hydraulic pressures within the braking system. This front/rear proportioning of braking forces will be optimal when it is in the exact same proportion as the ratio of the maximum braking forces and the front and rear axles ($F_{bf\,\text{max}}/F_{br\,\text{max}}$). Thus maximum braking forces (with tires at the point of an impending slide) will be developed when the brake force proportioning is

$$\frac{K_{bf}}{K_{br}} = \frac{F_{bf\,\text{max}}}{F_{br\,\text{max}}}$$

$$= \frac{l_2 + h(\mu + f_r)}{l_1 - h(\mu + f_r)} \tag{2.30}$$

Example 2.5

A new front-wheel-drive car has a wheelbase of 80 in. and a center of gravity that is 30 in. behind the front axle and at a height of 24 in. If the car is traveling at 80 mph, determine the proportion of total braking forces (front and rear axles) that will ensure that maximum braking forces of all tires will be developed at the same time on a concrete pavement with a coefficient of road adhesion of 0.6.

Solution

Using Eq. 2.30, we obtain

$$\frac{K_{bf}}{K_{br}} = \frac{50 + 24\left(0.6 + 0.01\left(1 + \dfrac{80 \times 1.47}{147}\right)\right)}{30 - 24\left(0.6 + 0.01\left(1 + \dfrac{80 \times 1.47}{147}\right)\right)}$$

or 81 percent of the braking force allocated to the front and 19 percent to the rear brakes.

Unfortunately, the design of a braking system is not as easy as it may seem at first glance, since optimal brake proportioning changes with vehicle and road

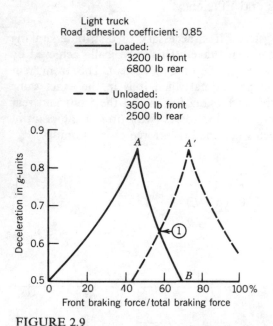

FIGURE 2.9
Effect of brake proportioning on the braking performance of a light truck. (Reproduced by permission of the Society of Automotive Engineers from D. J. Bickerstaff and G. Hartley, "Light Truck Tire Traction Properties and Their Effects on Braking Performance," *SAE Transactions*, Vol. 83, paper 741137, 1974.)

conditions. For example, the addition of vehicle cargo and/or passenger will change both the total weight and the distribution of weight, resulting in a new optimal brake proportioning. Similarly, changes in road conditions will produce different coefficients of adhesion again leading to revised optimal brake proportioning.

Figure 2.9 illustrates the effect of loading conditions on braking performance (for a light truck). The figure curves define the points of impending wheel lockup so, for example, with 20 percent of the braking force allocated to the front brakes, 0.6 g's will be achieved before the brakes lock and the coefficient of adhesion drops to slide values. Point A defines the optimal brake proportioning, as defined in Eq. 2.30, under loaded conditions as roughly 42 percent, of the total braking force allocated to the front brakes enabling a deceleration of 0.85 g's (i.e., equaling the coefficient of road adhesion, $\mu = 0.85$). Note that if only 20 percent of the braking force is allocated to the front brakes, an equivalent of 0.6 g's will be developed, which will result in substantially less than the optimal 0.85 g's. The point A' represents optimal proportioning under unloaded conditions as

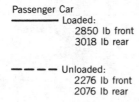

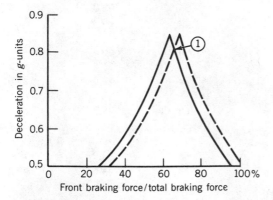

FIGURE 2.10
Effect of brake proportioning on the braking performance of a passenger car. (Reproduced by permission of the Society of Automottive Engineers from D. J. Bickerstaff and G. Hartley, "Light Truck Tire Traction Properties and Their Effects on Braking Performance", *SAE Transactions*, Vol. 83, paper 741137, 1974.)

70 percent of the braking force allocated to the front brakes. This disparity in optimal brake force allocation under loaded and unloaded conditions (points A and A') presents a dilemma that necessitates a compromised brake system design (e.g., the compromised brake proportioning represented by point 1) that gives suboptimal braking forces under both loaded and unloaded conditions. Figure 2.10 shows that the disparity of optimal braking forces between loaded and unloaded conditions is less pronounced for passenger cars but still necessitates brake system compromises.

Direct inspection of Eq. 2.30 indicates that in addition to loading conditions, differing coefficients of adhesion will result in differing optimal brake force distributions. Table 2.5 presents optimal values for actual passenger cars with coefficients of adhesion of 0.6 and 0.8. In examining this table, it is important to note that studies have indicated that if wheel lockup is to occur, it is preferable that the front wheels lock first, since rear wheel lockup can result in an uncontrollable vehicle spin. In contast, front-wheel lockup, while resulting in the

TABLE 2.5
Examples of Brake Proportioning for Selected Automobiles

Basic Parameters	Estimates of Early 1970s Vintage Cars		Actual Post 1980 Vintage Cars	
	Car 1	Car 2	Car 1	Car 2
Wheelbase, L (in.)	90	105	88.6	104.9
Unloaded				
weight, W (lb)	2000	3000	2185	3075
front weight (%)	60	54	60.0	62.4
center of gravity, h (in.)	20	20	19.0	21.0
Loaded				
weight, W (lb)	2600	3800	2660	3635
front weight (%)	51	47	52.4	54.2
center of gravity, h (in.)	20	20	18.0	21.0
Actual front brake proportioning (%)	60	60	79.7	80.6
Ideal front brake proportioning at $\mu = 0.6$				
unloaded, front (%)	73	65	72.9	74.4
loaded, front (%)	64	58	64.6	66.2
Ideal front brake proportioning at $\mu = 0.8$				
unloaded, front %	78	69	77.2	78.4
loaded, front %	69	62	68.7	70.2

Source: S. G. Shadle, L. H. Emery, and H. K. Brewer, "Vehicle Braking, Stability, and Control," *SAE Transactions*, Vol. 92, paper 830562, 1983.

loss of steering control, ensures that the vehicle will brake in a straight line [Campbell 1978; Wong 1978; and Malliaris et al. 1983]. With this in mind, Table 2.5 indicates that typical automobiles of the 1970s not only had lower braking efficiencies under most road conditions, but also were underbraked in the front relative to their 1980 vintage counterparts.

Since true optimal brake proportioning is seldom achieved because of necessary system compromises, it is useful to define a braking efficiency term that reflects the degree to which the braking system is operating below optimal. Simply stated, braking efficiency is defined as the ratio of the maximum rate of deceleration, expressed in g's (g_{max}), achievable prior to any wheel lockup to the coefficient of road adhesion

$$\eta_b = \frac{g_{max}}{\mu} \tag{2.31}$$

Example 2.6

Using Fig. 2.9, determine the braking efficiency of a loaded light truck on a road with a coefficient of adhesion of 0.85 and 40 percent of the braking force proportioned to the front axle.

Solution

By inspection of Fig. 2.9, the curve for loaded conditions indicates that when 40 percent of the braking force is allocated to the front, the deceleration rate is 0.75 g's. Using Eq. 2.31, we obtain

$$\eta_b = \frac{g_{max}}{\mu}$$

$$= \frac{0.75}{0.85}$$

$$= \underline{\underline{88.23}} \text{ percent}$$

2.9.3 Minimum Stopping Distance (Theoretical)

With a basic understanding of brake force proportioning and the resulting brake efficiency, attention can now be directed toward developing expressions for minimum stopping distances. By inspection of Fig. 2.8, the relationship between the stopping distance, braking force, vehicle mass, and vehicle speed is

$$a \, ds = \left[\frac{F_b + \Sigma R}{\gamma_b m} \right] ds$$

$$= V \, dV \tag{2.32}$$

Where γ_b is the mass factor accounting for the moments of inertia during braking, which for automobiles is taken as 1.04 [Wong 1978]. Integrating to determine stopping distance gives

$$S = \int_{V_1}^{V_2} \gamma_b m \frac{V \, dV}{F_b + \Sigma R} \tag{2.33}$$

Substituting in resistances (see Fig. 2.8), we obtain

$$S = \gamma_b m \int_{V_1}^{V_2} \frac{V \, dV}{F_b + f_r W \pm W \sin \theta_g + R_a} \tag{2.34}$$

where V_1 is the initial vehicle speed, V_2 is the final vehicle speed, and the grade resistance ($W \sin \theta_g$) is positive for uphill slopes and negative for downhill slopes. From Eq. 2.3, define an air resistance term (that does not vary with speed) $K_a = \rho / 2 C_D A_f$, so that

$$R_a = K_a V^2 \tag{2.35}$$

To simplify matters, assume that the effect of speed on the coefficient of rolling resistance, f_r, is constant and can be approximated by using the average of initial (V_1) and final (V_2) speeds in Eq. 2.6 (i.e., $V = (V_1 + V_2)/2$). With this assumption, and letting $m = W/g$, $F_b = \mu W$, the integration of Eq. 2.34 gives

$$S = \frac{\gamma_b W}{2 g K_a} \ln \left[\frac{\mu W + f_r W \pm W \sin \theta_g + K_a V_1^2}{\mu W + f_r W \pm W \sin \theta_g + K_a V_2^2} \right] \tag{2.36}$$

If the vehicle is assumed to stop ($V_2 = 0$),

$$S = \frac{\gamma_b W}{2 g K_a} \ln \left[1 + \frac{K_a V_1^2}{\mu W + f_r W \pm W \sin \theta_g} \right] \tag{2.37}$$

With braking efficiency considered, the actual braking force is

$$F_b = \eta_b \mu W \tag{2.38}$$

Therefore, by substitution, the theoretical stopping distance is

$$S = \frac{\gamma_b W}{2gK_a} \ln \left[1 + \frac{K_a V_1^2}{\eta_b \mu W + f_r W \pm W \sin \theta_g} \right] \tag{2.39}$$

Similarly, Eq. 2.36 can be rewritten to include braking efficiency. Finally, if aerodynamic resistance is ignored, deceleration is constant during braking and integration of Eq. 2.33 gives

$$S = \frac{\gamma_b (V_1^2 - V_2^2)}{2g(\eta_b \mu + f_r \pm \sin \theta_g)} \tag{2.40}$$

Example 2.7

A 2500-lb sports car is traveling at 90 mph ($C_D = 0.25$, $A_f = 18$ ft^2, $\rho = 0.0024$ slugs/ft^3) down a 10 percent grade. If the braking efficiency is 100 percent, the coefficient of road adhesion is 0.7, and the coefficient of rolling resistance is approximated as 0.019, determine the theoretical minimum stopping distance: (a) including aerodynamic resistance and (b) excluding aerodynamic resistance.

Solution

With aerodynamic resistance included, Eq. 2.39 can be applied with

$$\gamma_b = 1.04$$

$$\theta_g = 5.71°$$

$$K_a = \frac{0.0024}{2}(0.25)(18)$$

$$= 0.0054$$

Then

$$S = \frac{1.04(2500)}{64.4(0.0054)} \ln \left[1 + \frac{0.0054(90 \times 1.47)^2}{(1.0)(0.7)(2500) + (0.019)(2500) - 2500 \sin(5.71°)} \right]$$

$$= 442.88 \text{ ft}$$

With aerodynamic resistance excluded, Eq. 2.40 is used,

$$S = \frac{1.04(90 \times 1.47)^2}{64.4(0.7 + 0.019 - \sin(5.71°))}$$

$$= 456.64 \text{ ft}$$

Example 2.8

A car is traveling at 80 mph on a concrete surface and has an 80 percent braking efficiency. The brakes are applied in an attempt to miss an object that is 150 feet from the point of brake application and the coefficient of road adhesion is 0.85. Ignoring aerodynamic resistance and assuming theoretical minimum stopping distance, estimate how fast the car will be going when it strikes the object if (a) the surface is level, and (b) the surface is on a 5 percent uphill grade (noting that in both cases the coefficient of rolling resistance is approximated as 0.013).

Solution

For the both level and inclined surfaces, Eq. 2.40 is used, with the result that

$$S = \frac{\gamma_b(V_1^2 - V_2^2)}{2g(\eta_b\mu + f_r \pm \sin\theta_g)}$$

$$150 = \frac{1.04((80 \times 1.47)^2 - V_2^2)}{64.4\left(0.8(0.85) + 0.01\left(1 + \frac{40(1.47)}{147}\right) + 0\right)}$$

$$V_2 = 85.93 \text{ fps} \quad \text{or} \quad 58.45 \text{ mph}$$

On a 5 percent grade, $\theta_g = 2.86°$, we obtain

$$150 = \frac{1.04((80 \times 1.47)^2 - V_2^2)}{64.4\left(0.8(0.85) + 0.01\left(1 + \frac{40(1.47)}{147}\right) + 0.05\right)}$$

$$V_2 = 83.18 \text{ fps} \quad \text{or} \quad 56.59 \text{ mph}$$

2.9.4 Practical Stopping Distances

As previously mentioned, one of the most critical concerns in the design of highways is the provision of adequate driver sight distance to permit a safe stop. The theoretical assessment of stopping distance presented in the previous section provided the principles of braking for an individual vehicle under specified roadway surface conditions, but highway engineers face the problem of designing a roadway for a variety of driver skill levels, vehicle types, and weather conditions. Driving skills play an important roll in stopping-distance determination, since drivers that lock one or more wheels during emergency braking reduce the effective coefficient of road adhesion to slide values (see Table 2.4). Vehicles with antilock braking systems compensate for driver inexperience, but not all vehicles are so equipped [Rowell and Gritt 1982]. Next, the multitude of vehicle types gives a range of aerodynamics, weight distributions, tire conditions, and braking efficiencies that also must be accounted for in design. Finally, different weather conditions result in different coefficients of road adhesion (which are difficult to measure under any conditions due to nonuniform roadway surfaces) again introducing variability in vehicle stopping distances.

As a result of the wide variability inherent in the determination of braking distance, an equation that provides rough estimates for typical observed braking distances, and is more simplistic and usable than Eq. 2.39, is appropriate for design purposes. To begin deriving such an equation, note that if aerodynamic resistance, rolling resistance, moments of inertia, and braking efficiency are ignored, the summation of longitudinal vehicle forces in Fig. 2.8 can be written as

$$F_b \pm W \sin \theta_g = ma \qquad (2.41)$$

where the grade resistance ($W \sin \theta_g$) is positive for positive grades and negative for negative grades. As before, since highway grades are typically small, it is assumed that $\sin \theta_g = \tan \theta_g = G$, where G is the percent grade divided by 100 as previously defined.

Now define a friction term, f, such that $F_b = fW$. Solving Eq. 2.41 for acceleration, a, and substituting $F_b = fW$ and $m = W/g$ yields

$$a = -\left[\frac{g}{W}(fW \pm WG)\right] \quad \text{or} \quad a = -[g(f \pm G)]$$

with the minus sign denoting deceleration. Substitution into $V_2^2 = V_1^2 + 2ad$ [where d is the distance (ft), V_2 is the final speed (fps), V_1 is the initial speed (fps)] gives

$$d = \frac{V_1^2 - V_2^2}{2g(f \pm G)} \qquad (2.42)$$

If $V_2 = 0$ (i.e., the vehicle comes to a complete stop), the practical stopping-distance equation is

$$d = \frac{V_1^2}{2g(f \pm G)} \qquad (2.43)$$

It is important to note the similarity between Eq. 2.43 and Eq. 2.40 (the theoretical stopping distance ignoring aerodynamic resistance). Upon excluding

TABLE 2.6
Coefficients of Friction Used in Practical Stopping-Distance Computations

Initial Vehicle Speed (mph)	Coefficient of Friction (f)	Braking Distance on Level (ft)
20	0.40	33.3
25	0.38	54.8
30	0.35	85.7
35	0.34	120.1
40	0.32	166.7
45	0.31	217.7
50	0.30	277.8
55	0.30	336.1
60	0.29	413.8
65	0.29	485.6
70	0.28	593.3

Source: American Association of State Highway and Transportation Officials, "A Policy on Geometric Design of Highways and Streets, Washington, D.C., 1984.

the mass factor, ignoring the grade approximation ($\sin \theta_g = G$), and assuming that the wheels are at the point of impending slide, $f = \eta_b \mu + f_r$.

The f values used in Eq. 2.43 are determined from actual stopping experiments. With known initial speed and grade, test vehicles are brought to a stop, the required stopping distance is measured, and Eq. 2.43 is applied to solve for f. The values of f used in highway design are conservative estimates that are based on the assumption of some of the worst driver skills, roadway and tire conditions, and vehicle efficiencies likely to be encountered. Table 2.6 presents f values accepted for use in highway design.

It is important to note that the f values determined by the experimental method described above implicitly include the effects of aerodynamic resistance, braking efficiency, coefficient of road adhesion (with locked wheels), and inertia during braking (mass factor). Thus they are a function of the technology of the vehicles used in the experiments and engineers should recognize this when applying the practical stopping-distance equation (Eq. 2.43).

Example 2.9

A car ($C_D = 0.4$, $A_f = 28$ ft^2, $\rho = 0.0024$ slugs/ft^3, $W = 3000$ lb) is being used in a test on level concrete pavement to arrive at f values for stopping-distance equations. The coefficient of road adhesion is known to be 0.75 and the vehicle's braking efficiency is 0.85. If, during the test, the tires are at the point of impending skid and theoretical minimum stopping distance is achieved, what value of f will be arrived at in a stop from 60 mph?

Solution

First, for calculating the theoretical minimum stopping distance, Eq. 2.39 is applied,

$$S = \frac{\gamma_b W}{2gK_a} \ln\left[1 + \frac{K_a V_1^2}{\eta_b \mu W + f_r W \pm \sin\theta_g}\right]$$

$$K_a = \frac{\rho}{2} C_D A_f$$

$$= \frac{0.0024}{2}(0.4)(28)$$

$$= 0.0134$$

From Eq. 2.6, we obtain

$$f_r = 0.01\left(1 + \frac{(30 \times 1.47)}{147}\right)$$

$$= 0.013$$

so that

$$S = \frac{1.04(3000)}{64.4(0.0134)} \ln\left[1 + \frac{0.0134(60 \times 1.47)^2}{0.85(0.75)(3000) + 0.013(3000) + 0}\right]$$

$$S = 188.14 \text{ ft}$$

In solving for f, Eq. 2.43 is used,

$$d = \frac{V_1^2}{2gf}$$

$$f = \frac{V_1^2}{2gd}$$

$$= \frac{(60 \times 1.47)^2}{64.4(188.14)}$$

$$= 0.642$$

Example 2.10

Consider the test conditions in Example 2.9. If a new car ($C_D = 0.25$, $A_f = 18$ ft^2, $W = 2200$ lb) has a braking efficiency of 100 percent and is tested under the same conditions, how inaccurate will be the stopping distance predicted by the practical stopping-distance equation ($f = 0.642$)? How inaccurate will the stopping-distance prediction be if the new car has the same 85 percent braking efficiency as the original test car?

Solution

For the case with 100 percent braking efficiency,

$$f_r = 0.013 \qquad \text{as in Example 2.9}$$

$$K_a = \frac{\rho}{2} C_D A_f$$

$$= \frac{0.0024}{2}(0.25)(18)$$

$$= \underline{0.0054}$$

Using Eq. 2.39, we obtain

$$S = \frac{1.04(2200)}{64.4(0.0054)} \ln\left[1 + \frac{0.0054(60 \times 1.47)^2}{(1)(0.75)(2200) + 0.013(2200)}\right]$$

$$= 162.62$$

Now applying Eq. 2.43, we find that

$$d = \frac{V_1^2}{2gf}$$

$$= \frac{(60 \times 1.47)^2}{64.4(0.642)}$$

$$= 188.16$$

Therefore, the error is $\underline{25.54}$ ft, or 16 percent. By a similar procedure, the case with 85 percent braking efficiency,

$$S = \frac{1.04(2200)}{64.4(0.0054)} \ln\left[1 + \frac{0.0054(60 \times 1.47)^2}{0.85(0.75)(2200) + 0.013(2200)}\right]$$

$$= 190.34$$

Therefore, the error is $\underline{2.18}$ ft, or 1 percent.

Example 2.10 demonstrates the great reductions in stopping distance that can result from increased braking efficiency. Current antilock brake technology provides substantial reductions in stopping distances by preventing the coefficient of road adhesion from dropping to slide values, but it does not improve brake efficiency as defined in Eq. 2.31. The next major advance in brake technology will probably be a braking system that allocates brake force proportioning in response to changing vehicle weight distributions and coefficients of road adhesion.

2.9.5 Distance Traveled During Perception/Reaction

Until now, focus has been directed toward the distance required to stop a vehicle from the point of brake application. However, in providing a driver sufficient stopping-sight distance, it is also necessary to consider the distance traveled during the time in which the driver is perceiving and reacting to the need to stop. Thus the stopping-sight distance requirement is

$$d_{si} = d_r + d \tag{2.44}$$

and the distance traveled during perception/reaction is

$$d_r = V_1 t_r \tag{2.45}$$

where t_r is the time (in seconds) required to perceive and react to the need to stop.

The perception/reaction time of a driver is a function of a number of factors, including the driver's age, physical condition, and emotional state as well as the complexity of the situation and the strength of the stimuli requiring a stopping action. Numerous studies have provided valuable input on typical driver perception/reaction times [Johansson and Rumar 1971; T. Rockwell 1972]. From these studies, it has been determined that a conservative reaction time assumption of 2.5 sec is appropriate for highway design [American Association of State Highway and Transportation Officials 1984].

Example 2.11

Two drivers have standard reaction times (2.5 sec). One is obeying a 55-mph speed limit and the other is traveling illegally at 70 mph. How much distance will each of the two drivers cover during reaction time?

Solution

For driver one, traveling at 55 mph,

$$d_r = V_1 t_r = (55 \times 1.47)(2.5) = 202.125 \text{ ft}$$

For the driver traveling at 70 mph,

$$d_r = V_1 t_r = (70 \times 1.47)(2.5) = 257.25 \text{ ft}$$

Therefore, driving at 70 mph increases the distance covered by perception/reaction by 55.125 ft.

NOMENCLATURE
FOR
CHAPTER 2

a	acceleration (or deceleration if negative)
A_f	frontal area
C_D	drag coefficient
D_a	distance to accelerate
d	practical stopping distance
d_r	distance traveled during driver reaction
d_{si}	stopping-sight distance
F	total available tractive effort
F_b	total braking force
F_{bf}	front-axle braking force
$F_{bf\,max}$	maximum front-axle braking force
F_{br}	rear-axle braking force
$F_{br\,max}$	maximum rear-axle braking force
F_e	engine-generated tractive effort
F_f	available tractive effort at the front axle

F_{max}	maximum tractive effort
F_r	available tractive effort at the rear axle
f	friction term for practical stopping distance
f_r	coefficient of rolling resistance
G	percent grade ($G = 0.05$ is 5 percent grade)
g	gravitational constant
h	height of center of gravity above roadway surface
hp_{R_a}	horsepower required to overcome aerodynamic resistance
hp_{R_r}	horsepower required to overcome rolling resistance
i	running gear slippage
K_a	aerodynamic resistance constant
K_{bf}	proportion of braking force allocated to the front axle
K_{br}	proportion of braking force allocated to the rear axle
L	wheelbase
l_1	distance from the center of gravity to the front axle
l_2	distance from the center of gravity to the rear axle
M_e	engine torque output
m	mass
n_e	engine speed (revolutions per unit time)
P_{R_a}	power required to overcome aerodynamic resistance
R_a	aerodynamic resistance
R_g	grade resistance
R_r	total rolling resistance
R_{rf}	rolling resistance of the front axle
R_{rr}	rolling resistance of the rear axle
r	wheel radius
S	minimum theoretical stopping distance
t_r	driver reaction time
V	vehicle speed
W	total vehicle weight

W_f vehicle weight acting normal to the roadway surface on the front axle

W_r vehicle weight acting normal to the roadway surface on the rear axle

γ_b braking mass factor

γ_m acceleration mass factor

ε_0 gear reduction ratio

η_b braking efficiency

η_t gear reduction efficiency

θ_g angle of grade

μ coefficient of road adhesion

ρ air density

REFERENCES

1. C. Campbell, "The Sports Car: Its Design and Performance," Robert Bentley, Inc., Cambridge, Mass., 1978.

2. J. H. Wong, *Theory of Ground Vehicles*, John Wiley & Sons, New York, 1978.

3. H. K. Brewer and R. S. Rice, "Tires: Stability and Control," *SAE Transactions*, Vol. 92, paper 830561, 1983.

4. A. J. Scribor-Rylski, *Road Vehicle Aerodynamics*, John Wiley & Sons, 2nd ed., New York, 1984.

5. L. J. Janssen and W. H. Hucho, "The Effect of Various Parameters on the Aerodynamic Drag of Passenger Cars," *Advances in Road Vehicle Aerodynamics 1973*, BHRA Fluid Engineering, Cranfield, England.

6. J. J. Taborek, "Mechanics of Vehicles," *Machine Design*, 1957.

7. *Road & Track*, MGC Road Test, May 1969.

8. J. C. Hillard and G. Springer, eds., *Fuel Economy*, Plenum Press, New York, 1984.

9. Society of Automotive Engineers, "Engine Oil Effects on Vehicle Fuel Economy," PT-29, Warrendale, Pa., 1982.

10. Institution of Mechanical Engineers, "Driveline 84," I Mech E Conference Publications 1983-12, London, England, 1983.

11. S. G. Shadle, L. H. Emery, and H. K. Brewer, "Vehicle Braking, Stability and Control," *SAE Transactions*, Vol. 92, paper 830562, 1983.

12. D. J. Bickerstaff and G. Hartley, "Light Truck Tire Traction Properties and Their Effects on Braking Performance," *SAE Transactions*, Vol. 83, paper 741137, 1974.

13. A. C. Malliaris, R. Nicholson, J. Hedlund, and S. Scheiner, "Problems in Crash Avoidance and in Crash Avoidance Research," *SAE Transactions*, Vol. 92, paper 830560, 1983.

14. J. M. Rowell and P. S. Gritt, eds., Anti-Lock Braking Systems for Passenger Cars and Light Trucks—A Review," Society of Automotive Engineers, PT-27, Warrendale, Pa., 1982.

15. American Association of State Highway and Transportation Officials, "A Policy on Geometric Design of Highways and Streets," Washington, D.C., 1984.

16. G. Johansson and K. Rumar, "Drivers' Brake Reaction Times," *Human Factors*, Vol. 13, No. 1, 1971.

17. T. Rockwell, "Skills, Judgment, and Information Acquisition in Driving," in *Human Factors in Traffic Safety Research*, T. W. Forbes, ed., John Wiley & Sons, New York, 1972.

PROBLEMS

2.1. A new sports car has a drag coefficient of 0.29, a frontal area 18.5 ft^2, and is traveling at 100 mph. How much horsepower is required to overcome aerodynamic drag if $\rho = 0.002378$ slugs/ft^3?

2.2. A vehicle manufacturer is considering an engine for a new sedan ($C_D = 0.25$, $A_f = 17$ ft^2). The car will be tested at 100 mph maximum speed on a concrete paved surface at sea level ($\rho = 0.002378$ slugs/ft^3). The car currently weighs 2100 lb, but the designer selected an under-powered engine because he did not account for aerodynamic and rolling resistances. If 2 lb of additional vehicle weight is added for each unit of horsepower needed to overcome the neglected resistance, what will be the final weight of the car if it is to achieve its 100 mph speed?

2.3. For Example 2.3, how far back from the front axle would the center of graivty have to be to ensure that the maximum tractive effort developed for front- and rear-wheel drive options is equal (assume that all other variables are unchanged)?

2.4. A rear-wheel-drive 3000-lb drag race car has a 200-in. wheelbase and a center of gravity 20 in. above the pavement and 140 in. behind the front axle. The owners wish to achieve an initial acceleration from rest of 15 ft/sec^2 on a level concrete surface. What is the minimum coefficient of road adhesion needed to achieve this acceleration?

2.5. If the race car in Problem 2.4 had a center of gravity 70 in. above the ground and was run on a pavement with a coefficient of road adhesion of 0.95, how far back from the front axle would the center of gravity have to be to develop a maximum acceleration from rest of 1.2 g's (i.e., 38.64 ft/sec^2)?

2.6. A car is traveling on a concrete pavement with $C_D = 0.35$, $A_f = 21$ ft^2, $W = 3000$ lb, and $\rho = 0.002378$ slugs/ft^3; its engine is running at 3000 rpm and is producing 250 ft-lb of torque. If the gear reduction is ratio 3.5 to 1, gear reduction efficiency is 90 percent, running gear slippage is 3.5 percent, and wheel rolling radius is 15 in., determine the initial acceleration of the vehicle on a level road. (Assume that the available tractive effort is the engine-generated tractive effort.)

2.7. A car ($C_D = 0.35$, $A = 25$ ft^2, $\rho = 0.0024$ slugs/ft^3, and $W = 2500$ lb) has 14-in. radius wheels, a gear reduction efficiency of 90 percent, an overall gear reduction ratio of 3.2 to 1, and running gear slippage of 3.5 percent. The engine develops a maximum torque of 200 ft-lb at 3500 rpm. What is the maximum grade this vehicle could ascend, on a typical paved surface (e.g., concrete), while the engine is developing maximum torque? (Assume that the available tractive effort is the engine-generated tractive effort.)

2.8. A 2500-lb car has a maximum speed (at sea level, level terrain, paved surface) of 150 mph with 14-in. radius wheels, a gear reduction of 3 to 1, and a gear reduction efficiency of 90 percent. It is known that at the car's top speed the engine is producing 200 ft-lb of torque. If the car's frontal area is 25 ft^2, what is its drag coefficient?

2.9. A rear-wheel-drive car weighs 2500 lb, has an 80-in. wheelbase, a center of gravity 30 in. above the roadway surface and 30 in. behind the front axle, a gear reduction efficiency of 75 percent, 14-in. radius wheels, and an overall gear reduction of 6 to 1. The car's torque/revolutions per second (rps) curve is given by:

$$\text{torque} = 6 \text{ rps} - 0.045 \text{ rps}^2.$$

If the car is on a paved roadway surface with a coefficient of adhesion of 0.75, determine its maximum acceleration from rest.

2.10. Consider the car in Problem 2.9. If it is known that the car achieves maximum speed at an overall gear reduction ratio of 4 to 1 and, at this speed, its engine produces maximum horsepower, determine the maximum speed. (Assume that the running gear slippage is 3.5 percent.)

2.11. Consider the situation described in Example 2.4. If the vehicle is redesigned with 13-in. radius wheels (assume that the mass factor is unchanged) and a center of gravity located at the same height but at the midpoint of the wheelbase, determine the acceleration for front- and rear-wheel-drive options.

2.12. An engineer designs a rear-wheel-drive car (without an engine) that weighs 2000 lb and has a 100-in. wheelbase, gear reduction efficiency of 80 percent, 14-in. radius wheels, gear reduction ratio of 6 : 1 and a center of gravity that is 35 in. above the roadway surface and 55 in. behind the front axle. An engine that weighs 3 lb for each ft-lb of developed torque is to be placed in the front portion of the car. Calculations show that for every 20 lb of engine weight, the car's center of gravity moves 1 in. closer to the front axle (but stays at the same height above the roadway surface). If the car is starting from rest on a paved roadway with a coefficient of adhesion of 0.8, select an engine size (weight and associated torque) that will result in the highest possible available tractive effort. (Assume a level concrete surface.)

2.13. If the car in Example 2.7 had $C_D = 0.45$ and $A_f = 25$ ft^2, what would have been the difference in minimum theoretical stopping distances with and without aerodynamic resistance considered (all other factors the same as in Example 2.7)?

2.14. A vehicle ($C_D = 0.40$, $A_f = 28$ ft^2, $\rho = 0.0024$ slugs/ft^3, $W = 3500$ lb) is driven on a surface with a coefficient of adhesion 0.5 and the coefficient of rolling friction is approximated as 0.015. Assuming minimum theoretical stopping distances, if the vehicle comes to a stop in 250 ft after brake application on a level surface and has a braking efficiency of 0.78, what was its initial speed if (a) aerodynamic resistance is considered and (b) if aerodynamic resistance is ignored?

2.15. A level test track has a coefficient of road adhesion of 0.75 and a car being tested has a coefficient of rolling friction that is approximated as 0.018. The vehicle is tested unloaded and achieves the theoretical minimum stop in 200 ft (from brake application). The initial speed was 60 mph. Ignoring aerodynamic resistance, what is the unloaded braking efficiency?

2.16. A small truck is to be driven down a 4 percent grade at 70 mph. The coefficient of road adhesion is 0.95 and it is known that the braking efficiency is 80 percent when the truck is empty and it decreases by one percentage point for every 100 lb of cargo added. Ignoring aerodynamic resistance, if the driver wants the truck to be able to achieve a minimum theoretical stopping distance of 250 ft from the point of brake application, what is the maximum amount of cargo (in pounds) that can be carried?

2.17. Consider the conditions in Example 2.8. The car has $C_D = 0.5$, $A_f = 25$ ft^2, $W = 3500$ lb, $\rho = 0.0024$ slugs/ft^3, and a coefficient of rolling friction that is approximated as 0.018 for all speed conditions. If aerodynamic resistance is considered in stopping, estimate how fast the car will be going when it strikes the object with level and 5 percent grades (all other conditions, speed, etc., as descirbed in Example 2.8).

2.18. A transportation student is driving on a level road on a cold rainy night and sees a construction sign 600 ft ahead. The student strikes the sign at a speed of 35 mph. If the student insists that the 55 speed limit was not exceeded, what would be the associated reaction time (use the practical stopping-distance formula)?

2.19. Two cars are traveling on level terrain at 60 mph on a road with a coefficient of adhesion of 0.8. The driver of car 1 has a 2.5-sec reaction time and the driver of car 2 has a 2.0-sec reaction time. Both cars are traveling side by side and the drivers are able to stop their respective cars in the same distance from first seeing a roadway obstacle (i.e., reaction plus vehicle stopping distance). If the braking efficiency of car 2 is 0.75, determine the braking efficiency of car 1. (Assume minimum theoretical stopping distance and ignore aerodynamic resistance.)

2.20. A civil engineering student claims that a country road can be safely negotiated at 70 mph in rainy weather. Because of the winding nature of the road, one stretch of level pavement has a sight distance of only 600 ft. Using the practical stopping-distance formula, comment on the student's claim.

2.21. A driver is traveling at 55 mph on a wet road. An object is spotted on the road 450 ft ahead and the driver stops just in time. Assuming standard reaction time and using the practical stopping-distance formula, determine the grade of the road.

2.22. A test on a driver's reaction time is being conducted on special testing track with a wet pavement (i.e., f-values in Table 2.6 apply) and a driving speed of 55 mph. When the driver is sober, a stop can be made just in time to avoid hitting an object that is first visible from 520 ft. After a few drinks under the exact same conditions, the driver fails to stop in time and strikes the object at a speed of 34 mph. Determine the driver's reaction time before and after drinking. (Use the practical stopping-distance formula.)

Chapter Three

Geometric Alignment of Highways

3.1 INTRODUCTION

With a basic understanding of vehicle performance in hand (from Chapter 2), attention can now be directed toward the manner in which such an understanding is used in the actual design of highways. The design of a highway necessitates the determination of specific design elements, which include items such as the number of lanes, lane width, median type and width, length of freeway acceleration and deceleration lanes, need for truck climbing lanes for highways on steep grades, and radii required for vehicle turning. For virtually all of these highway design elements, the performance characteristics of vehicles play an important role. For example, vehicle acceleration and deceleration characteristics have a direct impact on the design of acceleration and deceleration lanes (i.e., the length needed to provide a safe and orderly flow of traffic). Furthermore, they determine the need for truck climbing lanes on steep grades as well as the number of highway lanes required, since the spacing between vehicles is directly related to vehicle performance characteristics (this will be discussed further in Chapter 5). In addition, the physical dimensions of vehicles affect a number of design elements such as the radii required for turning, height of highway overpasses, lane widths, and so on.

When one considers the diversity of vehicles, in terms of performance and physical dimensions, in conjunction with the many elements comprising the design of a highway, it is clear that proper highway design is a complex procedure and one that requires numerous compromises. Moreover, design standards must change over time in response to changes in vehicle performance and dimensions as well as in response to physical evidence gathered on the effectiveness of existing highway designs. Current standards of highway design are presented in detail in "A Policy on Geometric Design of Highways and Streets 1984," published by the American Association of State Highway and Transportation Officials [AASHTO 1984].

Because of the shear number of geometric elements involved in the design of a highway, presenting a detailed discussion of each design element is well beyond the scope of this book and the reader is referred to other sources for such information [Ibid; Wright and Paquette 1987]. Instead, this book focuses on a single, highly significant highway geometric concern: highway alignment. As will be shown, the alignment topic is particularly well suited for demonstrating the effect of vehicle performance (specifically braking performance) and vehicle dimensions (e.g. height of the driver's eye and headlight height) on the design of highways. By concentrating on the specifics of the highway alignment problem, the student will develop a flavor for the procedures and compromises inherent in the design of all highway geometric elements.

3.2 PRINCIPLES OF HIGHWAY ALIGNMENT

The alignment of a highway is a three-dimensional problem with measurement in x, y, and z dimensions. This is illustrated, from a driver's perspective, in Fig. 3.1

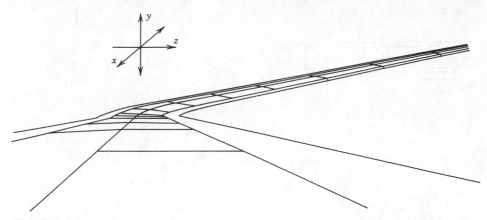

FIGURE 3.1
Highway alignment in 3-dimensions. (Reproduced by permission from F. L. Mannering, Computer Plotting of Highway Perspectives, unpublished Bachelor's Thesis, University of Saskatchewan, Saskatoon, Canada, 1976.)

[Mannering 1976]. However, in highway design practice, three-dimensional design computations are cumbersome and, what is perhaps more important, the actual implementation and construction of a design based on three-dimensional coordinates has historically been beyond the practical capabilities of existing technology. As a consequence, the three-dimensional highway alignment problem is typically reduced to 2 two-dimensional alignment problems, as illustrated in Fig. 3.2. One of the alignment problems in this figure corresponds roughly to x and z coordinates and is referred to as horizontal alignment. The other corresponds to highway length and y coordinates (i.e., elevation). Referring to Fig. 3.2, note that the horizontal alignment of a highway is represented in a *plan view*, which is essentially equivalent to the perspective of an aerial photo of the highway. The vertical alignment is represented in a *profile view*, which gives the elevation of all points along the length of the highway.

Aside from considering the alignment problem as 2 two-dimensional problems, one further simplification is made. That is, instead of using x and z coordinates, highway positioning and length is defined as the actual distance along the highway (usually measured along the centerline of the highway, on a horizontal plane) from some specified point. This distance is generally measured in terms of stations, with each station constituting 100 ft of highway alignment distance. The notation for stationing distance is such that a point on a highway 1258.5 ft from a specified point of origin is said to be at stationing 12 + 58.5 (i.e., 12 stations and 58.5 ft) with the point of origin being at stationing 0 + 00. This stationing concept, when combined with the alignment direction given in the plan view (horizontal alignment) and the elevation corresponding to stations given in the profile view (vertical alignment), gives a unique identification of all highway points in a manner virtually equivalent to using the true x, y, and z coordinates.

Plan View (horizontal alignment)

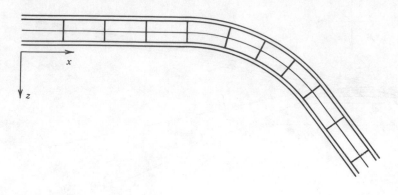

Profile View (vertical alignment)

FIGURE 3.2
Highway alignment in two-dimensional views.

3.3 VERTICAL ALIGNMENT

The objective of vertical alignment is to determine the elevation of highway points to ensure proper roadway drainage and an acceptable level of safety. The primary challenge of vertical alignment lies in the transition of roadway elevation between two grades. This transition is achieved by means of a vertical curve.

Vertical curves can be broadly classified into crest vertical curves and sag vertical curves as illustrated in Fig. 3.3 [AASHTO 1984]. In this figure, G_1 is the initial roadway grade (also referred to as the initial tangent grade, viewing Fig. 3.3 from left to right), G_2 is the final roadway (tangent) grade, A is the absolute value of the difference in grades (generally expressed in percent), L is the length of the vertical curve measured in a horizontal plane, PVC is the initial point of the vertical curve, PVI is the point of vertical intersection (intersection of initial and final grades), and PVT is the final point of the vertical curve. In practice, vertical curves are almost always arranged such that half of the curve length is positioned before the PVI and half after (as illustrated in Fig. 3.3). Curves that satisfy this criterion are referred to as equal tangent vertical curves.

In terms of referencing points on a vertical curve, it is important to note that the profile views presented in Fig. 3.3 correspond to all highway points even if a horizontal curve occurs concurrently with a vertical curve (as in Fig. 3.1). Thus

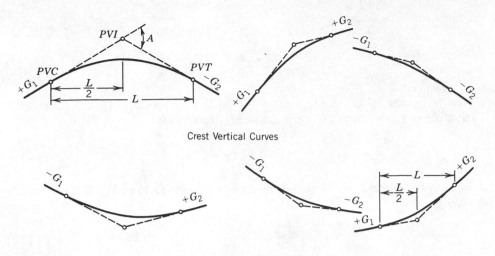

Crest Vertical Curves

Sag Vertical Curves

FIGURE 3.3
Types of vertical curves. (Reproduced by permission from American Association of State Highway and Transportation Officials, A Policy on Geometric Design of Highways and Streets, Washington D.C., 1984.)

each highway point is uniquely defined, vertically, by stationing (which is measured along a horizontal plane for vertical referencing) and elevation. This will be made clearer, through examples, in the following discussion.

3.3.1 Vertical Curve Fundamentals

In connecting roadway tangent grades with an appropriate vertical curve, a mathematical relationship defining roadway elevations at every point (or equivalently, station) is needed. A parabolic function has been found suitable in this regard because, among other things, it provides a constant rate of change of slope and implies equal curve tangents. The general form of the parabolic equation, as applied to vertical curves, is

$$y = ax^2 + bx + c \tag{3.1}$$

where y is the roadway elevation x stations (or feet) from the beginning of the vertical curve (i.e., the PVC). By definition, c is the elevation of the PVC, since $x = 0$ corresponds to the PVC. For defining a and b, note that the first derivative of Eq. 3.1 gives the slope and is

$$\frac{dy}{dx} = 2ax + b \tag{3.2}$$

At the PVC, $x = 0$, so that

$$b = \frac{dy}{dx}$$
$$= G_1 \tag{3.3}$$

where G_1 is the initial slope as previously defined. Also note that the second derivative of Eq. 3.1 is the rate of change of slope and is

$$\frac{d^2y}{dx} = 2a \tag{3.4}$$

However, the average rate of change of slope, by observation of Fig. 3.3, can also be written as

$$\frac{d^2y}{dx} = \frac{G_2 - G_1}{L} \tag{3.5}$$

Equating Eqs. 3.4 and 3.5 gives

$$a = \frac{G_2 - G_1}{2L} \tag{3.6}$$

Example 3.1

A 600-ft equal tangent sag vertical curve has the PVC at station $170 + 00$ and elevation 1000 ft. The initial grade is -3.5 percent and the final grade is 0.5 percent. Determine the elevation and stationing of the PVI, PVT, and the lowest point on the curve.

Solution

Since the curve is equal tangent, the PVI will be 300 ft or three stations (measured in a horizontal, profile plane) from the PVC, and the PVT will be 600 ft or six stations from the PVC. Therefore, the stationing of PVI and PVT are $173 + 00$ and $176 + 00$, respectively. For the elevations of the PVI and PVT, it is known that a -3.5 percent grade can be equivalently written as -3.5 ft/station (since percent grade is defined as vertical feet per 100 horizontal feet). Since the PVI is three stations from the PVC, which is known to be at elevation 1000 ft, the elevation of the PVI is

$$1000 - 3.5 \text{ ft/station (3 stations)} = 989.5 \text{ ft}$$

Similarly, with the PVI at elevation 989.5 ft, the elevation of the PVT is

$$989.5 + 0.5 \text{ ft/station (3 stations)} = 991.0 \text{ ft}$$

It is obvious from the values of the initial and final grades that the lowest point on the vertical curve will occur when the first derivative of the parabolic function (Eq. 3.1) vanishes. Note that this may not always be the case. For example, a sag curve with an initial grade of -2.0 percent and a final grade of -1.0 percent will have its lowest elevation point at the PVT and the first derivative will not vanish. However, in this example the deriviative will vanish:

$$\frac{dy}{dx} = 2ax + b$$

$$= 0$$

From Eq. 3.3, we obtain

$$b = G_1$$

$$= -3.5$$

and from Eq. 3.6,

$$a = \frac{0.5 - (-3.5)}{2(6)}$$

$$= 0.33$$

Substituting for a and b gives

$$\frac{dy}{dx} = 2(0.33)x - 3.5$$

$$= 0$$

$$x = 5.3 \text{ stations}$$

This gives the stationing of the low point at $175 + 30$ (i.e., $5 + 30$ stations from the PVC). For the elevation of the lowest point on the vertical curve, the values of a, b, c (elevation of the PVC), and x are substituted into Eq. 3.1 giving

$$y = 0.33(5.3)^2 + (-3.5)(5.3) + 1000$$

$$= 990.72 \text{ ft}$$

Note that the above equations can also be solved with grades expressed as the decimal equivalent of percent (e.g., 0.02 for 2 percent) if x is expressed in feet instead of stations.

Some additional properties of vertical curves can now be formalized. For example, offsets, which are vertical distances from the initial tangent to the curve as illustrated in Fig. 3.4, are extremely important for vertical curve design and construction. In Fig. 3.4, Y is the offset at any distance, x, from the PVC; Y_m is

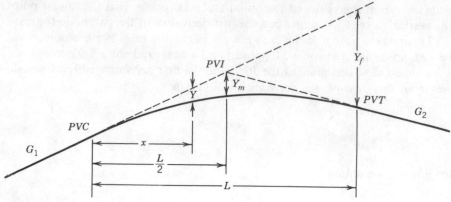

FIGURE 3.4
Offsets for vertical curves.

the midcurve offset; and Y_f is the offset at the end of the vertical curve. From the properties of an equal tangent parabola, it can be readily shown that

$$Y = \frac{A}{200L} x^2 \tag{3.7}$$

where Y is the offset in feet, A is the absolute value of the difference in grades ($G_2 - G_1$ expressed in percent), L is the length of the vertical curve in feet, and x is the distance from the *PVC* in feet. It follows from Fig. 3.4 that

$$Y_m = \frac{AL}{800} \tag{3.8}$$

and

$$Y_f = \frac{AL}{200} \tag{3.9}$$

Another useful vertical curve property is one that simplifes the computation of the high and low points of crest and sag vertical curves, respectively (given that the high or low point does not occur at the curve ends, *PVC* or *PVT*). Recall that in Example 3.1, the first derivative was used to determine the location of the low point. The alternative to this is to use a *K*-value defined as (with L in feet and A in percent),

$$K = \frac{L}{A} \tag{3.10}$$

The *K*-value can be used directly to compute the high/low points for crest/sag vertical curves by

$$x = KG_1 \tag{3.11}$$

where x is the distance from the PVC to the high/low point. In words, the K-value is the horizontal distance, in feet, required to affect a 1 percent change in the slope. Aside from the low/high point computations, it will be shown in Sections 3.3.3 and 3.3.4 that the K-value has many important applications in the design of vertical curves.

3.3.2 Minimum and Desirable Stopping-Sight Distances

Construction of a vertical curve is generally a costly operation requiring the movement of significant amounts of earthen material. Thus one of the primary challenges facing highway designers is to minimize construction costs while still providing an adequate level of safety. An appropriate level of safety is generally defined as that level of safety that provides drivers with sufficient sight distance to enable them to safely stop their vehicles in response to objects obstructing their forward motion (the provision of adequate roadway drainage is often an important concern too, but it is not discussed in this book; see [AASHTO 1984]. Referring back to the vehicle braking performance concepts discussed in Chapter 2, we can compute this necessary stopping sight distance (SSD) simply as the summation of the vehicle stopping distance (Eq. 2.43) and the distance traveled during perception/reaction time (Eq. 2.45). That is,

$$\text{SSD} = \frac{V_1^2}{2g(f \pm G)} + V_1 t_r \tag{3.12}$$

where SSD is the stopping-sight distance, V_1 is the initial vehicle speed, g is the gravitational constant, f is the coefficient of braking friction, G is the grade, and t_r is the perception/reaction time.

Given Eq. 3.12, the question now becomes one of selecting appropriate values for the computation of SSD. Values of the coefficient of braking friction term, f, are selected so as to be representative of poor driver skills, low braking efficiencies, and wet pavements, thus providing conservative or near worst-case stopping-sight distances (see Table 2.6 for such examples of f). Also, a standard perception/reaction time of 2.5 sec is typically used as a conservative estimate. In terms of the assumed initial vehicle speeds used in Eq. 3.12, two values are worthy of note. The first value is the design speed of the highway, which is defined as the maximum safe speed that a highway can be negotiated assuming near worst-case conditions. The second value is the average vehicle running speed, which is obtained from actual vehicle speed observations, and usually ranges from 90 to 95 percent of the highway's design speed. This lower speed reflects the fact that drivers generally drive below the design speed of the highway even under low-volume, uncongested traffic conditions. The application of Eq. 3.12 (with $G = 0$) produces stopping-sight distances presented in Table 3.1 for average running speeds (the lower of the assumed speeds for conditions) and design speeds. Obviously, since average running speeds are equal to or lower than corresponding design speeds (depending on the design speed value), the stopping-sight distance required will be generally lower for average running

TABLE 3.1
Stopping-Sight Distance (minimum and desirable)

Design Speed (mph)	Assumed Speed for Condition (mph)	Brake Reaction Time (sec)	Brake Reaction Distance (ft)	Coefficient of Friction f	Braking Distance on Level[a] (ft)	Stopping-Sight Distance Computed[a] (ft)	Stopping-Sight Distance Rounded for Design (ft)
20	20–20	2.5	73.3– 73.3	0.40	33.3– 33.3	106.7–106.7	125–125
25	24–25	2.5	88.0– 91.7	0.38	50.5– 54.8	138.5–146.5	150–150
30	28–30	2.5	102.7–110.0	0.35	74.7– 85.7	177.3–195.7	200–200
35	32–35	2.5	117.3–128.3	0.34	100.4–120.1	217.7–248.4	225–250
40	36–40	2.5	132.0–146.7	0.32	135.0–166.7	267.0–313.3	275–325
45	40–45	2.5	146.7–165.0	0.31	172.0–217.7	318.7–382.7	325–400
50	44–50	2.5	161.3–183.3	0.30	215.1–277.8	376.4–461.1	400–475
55	48–55	2.5	176.0–201.7	0.30	256.0–336.1	432.0–537.8	450–550
60	52–60	2.5	190.7–220.0	0.29	310.8–413.8	501.5–633.8	525–650
65	55–65	2.5	201.7–238.3	0.29	347.7–485.6	549.4–724.0	550–725
70	58–70	2.5	212.7–256.7	0.28	400.5–583.3	613.1–840.0	625–850

Source: American Association of State Highway and Transportation Officials, "A Policy on Geometric Design of Highways and Streets," Washington, D.C., 1984.

[a] Different values for the same speed result from using unequal coefficients of friction.

speeds. Stopping-sight distances derived from average running speeds are referred to as minimum SSDs, whereas stopping-sight distances derived from design speeds are referred to as desirable SSDs.

3.3.3 Stopping-Sight Distance and Crest Vertical Curve Design

In providing sufficient SSD on a vertical curve, the length of curve (L in Fig. 3.3) is the critical concern. Longer lengths of curve provide more SSD, all else being equal, but are more costly to construct. Shorter curve lengths are relatively inexpensive to construct but may not provide adequate SSD. What is needed, then, is an expression for minimum curve length given required SSD. In developing such an expression, crest and sag vertical curves are considered separately.

For the crest vertical curve case, consider the diagram presented in Fig. 3.5. In this figure, S is the sight distance, L is the length of curve, H_1 is the driver's eye height, H_2 is the height of a roadway object, and other terms are as previously defined. Using the properties of a parabola for an equal tangent curve, it can be shown that the minimum length of curve, L_m, for a required sight distance, S is

$$L_m = 2S - \frac{200(\sqrt{H_1} + \sqrt{H_2})^2}{A} \qquad \text{for } S > L \qquad (3.13)$$

$$L_m = \frac{AS^2}{200(\sqrt{H_1} + \sqrt{H_2})^2} \qquad \text{for } S < L \qquad (3.14)$$

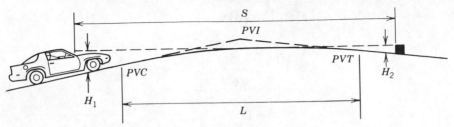

FIGURE 3.5
Stopping-sight distance considerations for crest vertical curves.

For the sight distance required to provide adequate SSD, current AASHTO design standards [1984] use a driver eye height, H_1, of 3.5 ft and a roadway object height, H_2, of 0.5 ft. It is also important to note that, when determining the sight distance, S is assumed to be equal to SSD. Substituting AASHTO standards for H_1 and H_2, and $S = $ SSD into Eqs. 3.13 and 3.14 gives

$$L_m = 2\,\text{SSD} - \frac{1329}{A} \qquad \text{for SSD} > L \qquad (3.15)$$

$$L_m = \frac{A\,\text{SSD}^2}{1329} \qquad \text{for SSD} < L \qquad (3.16)$$

Example 3.2

A highway is being designed to AASHTO standards with a 70-mph design speed and, at one section, an equal tangent vertical curve must be designed to connect grades of +1.0 percent and −3.0 percent. Determine the minimum length of vertical curve required assuming provisions are to be made for minimum SSD and desirable SSD.

Solution

For minimum SSD, the assumed average running speed is 58 mph from Table 3.1. If the conservative worst-case value of G is used (i.e., −3.0 percent) the SSD is computed from Eq. 3.12 as (using values from Table 3.1),

$$\text{SSD} = \frac{V_1^2}{2g(f \pm G)} + V_1 t_r$$

$$= \frac{(58 \times 1.47)^2}{2(32.2)(0.28 - 0.03)} + (58 \times 1.47)(2.5)$$

$$= 664.66 \text{ ft}$$

If we assume that SSD < L, Eq. 3.16 gives

$$L_m = \frac{A \, SSD^2}{1329}$$

$$= \frac{4(664.66)^2}{1329}$$

$$= 1329.64 \text{ ft}$$

Since 1329.64 > 664.66, the assumption that SSD < L is valid.

For desirable SSD, the initial vehicle speed is assumed to be equal to the highway's design speed. Again applying Eq. 3.12 with values from Table 3.1, we obtain

$$SSD = \frac{(70 \times 1.47)^2}{2(32.2)(0.28 - 0.03)} + (70 \times 1.47)(2.5)$$

$$= 914.92 \text{ ft}$$

If we assume that SSD < L, Eq. 3.16 gives

$$L_m = \frac{4(914.92)^2}{1329}$$

$$= 2519.42 \text{ ft}$$

Again the assumption that SSD < L is satisfied.

Working with Eqs. 3.15 and 3.16 can be cumbersome, particularly since an initial assumption regarding SSD > L or SSD < L must be made. To simplify matters on crest vertical curve computations (and also sag vertical curve computations as will be shown in the next section), K-values, as described earlier in Eqs. 3.10 and 3.11, are used. To compute K-values, note that Eq. 3.16 (for SSD < L) can be written as

$$L = KA \tag{3.17}$$

where again K is the horizontal distance, in feet, required to affect a 1 percent change in the slope. Given design SSDs, values of K can be readily computed. In a similar manner K-values can also be computed for the case where L > SSD. Table 3.2 presents these K-value computations, for crest vertical curves, for minimum and desirable stopping-sight distances.

In using the K-values in Table 3.2, two points are worthy of note. First, the SSDs used in the K-value computations assume that G = 0. In most vertical curve instances, this is not the conservative near worst-case assumption. For

TABLE 3.2
Design Controls for Crest Vertical Curves Based on Minimum and Desirable
Stopping-Sight Distance

Design Speed (mph)	Assumed Speed for Condition (mph)	Coefficient of Friction f	Stopping-Sight Distance, Rounded for Design (ft)	Rate of Vertical Curvature, K^a [length (ft) per percent of A]	
				Computed[b]	Rounded for Design
20	20–20	0.40	125–125	8.6– 8.6	10– 10
25	24–25	0.38	150–150	14.4– 16.1	20– 20
30	28–30	0.35	200–200	23.7– 28.8	30– 30
35	32–35	0.34	225–250	35.7– 46.4	40– 50
40	36–40	0.32	275–325	53.6– 73.9	60– 80
45	40–45	0.31	325–400	76.4–110.2	80–120
50	44–50	0.30	400–475	106.6–160.0	110–160
55	48–55	0.30	450–550	140.4–217.6	150–220
60	52–60	0.29	525–650	180.2–302.2	190–310
65	55–65	0.29	550–725	227.1–394.3	230–400
70	58–70	0.28	625–850	282.8–530.9	290–540

Source: American Association of State Highway and Transportation Officials, "A Policy on Geometric Design of Highways and Streets," Washington, D.C., 1984.

[a] Different K-values for the same speed result from using unequal coefficients of friction.

[b] Using computed values of stopping-sight distance.

instance, in Example 3.2, $G = -3.0$ percent was used in the SSD computation to provide a conservative estimate. If $G = 0$ percent had been assumed in this problem, the vertical curve lengths would have been 1143.11 ft and 2146.26 ft for the minimum and desirable designs, respectively, which are substantially less than the values computed with the conservative $G = -3.0$ percent assumption. In reality, depending on the initial and final grades of the curve, the true solution to the problem generally lies somewhere between the assumption of $G = 0$ percent and $G = -3.0$ percent. In practice, depending on jurisdictional district, policies as to how this problem is handled, vary. However, for the remainder of this book, the $G = 0$ percent option will be used, thus making the K-values presented in Table 3.2 valid for all of the design problems contained herein regardless of the percent of initial and final grades.

The second point is that low values of A can give curve lengths that are unrealistically low (or even negative if Eq. 3.15 is used [i.e., SSD > L]). As a result, it is common practice to set minimum curve length limits that range from 100 to 300 ft depending on individual jurisdictional standards. A common altenative is to set minimum curve lengths at three times the design speed [AASHTO 1984].

Example 3.3

Solve Example 3.2 using the K-values listed in Table 3.2.

Solution

From Example 3.2, $A = 4$. For minimum SSD, $K = 290$ from (Table 3.2) at a 70-mph design speed. Therefore, application of Eq. 3.17 gives

$$L_m = KA$$

$$= 290(4)$$

$$= \underline{\underline{1160 \text{ ft}}}$$

which is close (with the difference being the rounded K-values) to the 1143.11 ft that would have been computed in Example 3.2 had $G = 0$ percent been used in SSD computations.

For desirable SSD, $K = 540$ from (Table 3.2) at a 70-mph design speed, giving

$$L_m = 540(4)$$

$$= \underline{\underline{2160 \text{ ft}}}$$

which is approximately equal to the 2146.26 ft that would have been computed in Example 3.2 had $G = 0$ percent been used in SSD computations.

Example 3.4

If the stationing of the PVI for the curves connecting the grades in Example 3.2 is arbitrarily set at $100 + 00$ whether considering minimum or desirable SSD curve lengths, determine the stationing of the PVC, PVT, and curve high points for both minimum and desirable curve designs.

Solution

For minimum curve design, $L = 1160$ ft from Example 3.3. Since the curve is equal tangent (as are virtually all curves used in practice), one-half of the curve will occur before the *PVI* and one-half after, so that

$$PVC \text{ is at } 100 + 00 - \frac{L}{2} = 100 + 00 - 5 + 80$$

or at station 94 + 20

$$PVT \text{ is at } 100 + 00 + \frac{L}{2} = 100 + 00 + 5 + 80$$

or at station 105 + 80

For the stationing of the highpoint, Eq. 3.11 is used,

$$x = KG_1$$
$$= 290(1)$$
$$= 290 \text{ ft}$$

or at station, 94 + 20 + 2 + 90 = station 97 + 10

Similarly, for desirable curve design, the *PVC* can be shown to be at station 89 + 20, *PVT* at station 110 + 80 and the highpoint at station 94 + 60.

3.3.4 Stopping-Sight Distance and Sag Vertical Curve Design

Sag vertical curve design differs from crest vertical curve design in the sense that sight distance is governed by nighttime conditions because, in daylight, sight distance on a sag vertical curve is unresticted. Thus the critical concern for sag vertical curves is the headlight sight distance (i.e., the length of road illuminated by the vehicle's headlights) which is a function of the height of the headlight above the roadway, H, and the inclined upward angle of the headlight beam, relative to the horizontal plane of the car, β. The sag vertical curve sight distance problem is illustrated in Fig. 3.6. By using the properties of a parabola for an equal tangent curve (as was done for the crest vertical curve case), it can be shown that the minimum length of curve, L_m, for a required sight distance is

$$L_m = 2S - \frac{200(H + S \tan \beta)}{A} \qquad \text{for } S > L \qquad (3.18)$$

$$L_m = \frac{AS^2}{200(H + S \tan \beta)} \qquad \text{for } S < L \qquad (3.19)$$

For the sight distance required to provide adequate SSD, current AASHTO design standards [1984] use a headlight height of 2.0 ft and an upward angle of 1 degree. Substituting these design standards and $S = $ SSD (as was done for the

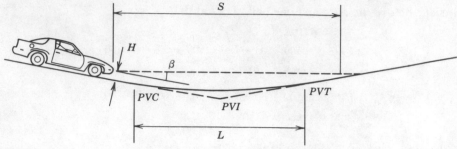

FIGURE 3.6
Stopping-sight distance considerations for sag vertical curves.

crest curve case) into Eqs. 3.18 and 3.19 gives

$$L_m = 2\,\text{SSD} - \frac{400 + 3.5\,\text{SSD}}{A} \qquad \text{for SSD} > L \qquad (3.20)$$

$$L_m = \frac{A\,\text{SSD}^2}{400 + 3.5\,\text{SSD}} \qquad \text{for SSD} < L \qquad (3.21)$$

As was the case for crest vertical curves, K-values can also be computed for sag vertical curves as listed in Table 3.3. Again, caution should be exercised in using the K-values in this table, since the assumption of $G = 0$ percent is used for SSD computations. Finally, minimum lengths for sag vertical curves generally follow the same limits set for crest vertical curves, as previously discussed in the paragraph preceding Example 3.3.

Example 3.5

An engineering mistake has resulted in the need to connect an already constructed tunnel and bridge with a vertical curve. The profile of the tunnel and bridge is given in Fig. 3.7. Devise a vertical alignment to connect the tunnel and bridge by determining the highest common design speed for the sag and crest (equal tangent) vertical curves needed. Assume that the final grade of the sag curve equals the initial grade of the crest curve. Compute stationing of PVC, PVI, and PVT curve points.

TABLE 3.3
Design Controls for Sag Vertical Curves Based on Minimum and Desirable
Stopping-Sight Distance

Design Speed (mph)	Assumed Speed for Condition (mph)	Coefficient of Friction f	Stopping-Sight Distance, Rounded for Design (ft)	Rate of Vertical Curvature, K^a [length (ft) per percent of A]	
				Computed[b]	Rounded for Design
20	20–20	0.40	125–125	14.7– 14.7	20– 20
25	24–25	0.38	150–150	21.7– 23.5	30– 30
30	28–30	0.35	200–200	30.8– 35.3	40– 40
35	32–35	0.34	225–250	40.8– 48.6	50– 50
40	36–40	0.32	275–325	53.4– 65.6	60– 70
45	40–45	0.31	325–400	67.0– 84.2	70– 90
50	44–50	0.30	400–475	82.5–105.6	90–110
55	48–55	0.30	450–550	97.6–126.7	100–130
60	52–60	0.29	525–650	116.7–153.4	120–160
65	55–65	0.29	550–725	129.9–178.6	130–180
70	58–70	0.28	625–850	147.7–211.3	150–220

Source: American Association of State Highway and Transportation Officals, "A Policy on Geometric Design of Highways and Streets," Washington, D.C., 1984.

[a]Different K-values for the same speed result from using unequal coefficients of friction.

[b]Using computed values of stopping-sight distance.

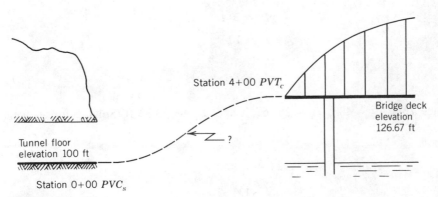

Station 4+00 PVT_c

Bridge deck elevation 126.67 ft

Tunnel floor elevation 100 ft

?

Station 0+00 PVC_s

FIGURE 3.7
Profile view of vertical alignment diagram for Example 3.5.

Solution

From left to right, a sag vertical curve (with subscripting s) and a crest vertical curve (with subscripting c) is needed to connect the tunnel and bridge. From given information, it is known that $G_{1s} = 0$ percent (i.e., the initial slope for the sag curve) and $G_{2c} = 0$ percent (i.e., the final slope for the crest vertical curve). The most efficient vertical design will use all of the horizontal distance in the profile view. This implies that the most efficient design will! have the PVT of the sag surve (PVT_s) be the PVC of the crest curve (PVC_s). Further the problem states that $G_{2s} = G_{1c}$ and since $G_{1s} = G_{2c} = 0$, $A_s = A_c = A$, the common algebraic difference in the grades.

Since 400 feet are provided,

$$L_s + L_c = 400$$

Also, the summation of the end-of-curve offset for the sag curve and the beginning-of-curve offset (relative to the final grade) for the crest vertical curve must equal 26.67 ft. From Eq. 3.9, this gives

$$\frac{AL_s}{200} + \frac{AL_c}{200} = 26.67$$

With two equations and three unknowns, a third equation is needed. Such an equation can be obtained by trial and error by using K-values for assumed design speeds. To begin, assume a 20-mph design speed. The corresponding K-values are 20 for the sag curve and 10 for the crest curve, as obtained from Tables 3.3 and 3.2, respectively. Use of Eq. 3.17 gives

$$L_s = 20A$$
$$L_c = 10A$$

or

$$L_s = 2L_c$$

and since $L_s + L_c = 400$, $L_s = 266.67$ ft and $L_c = 133.33$ ft. Solving for A, we obtain

$$A = \frac{L_c}{10}$$
$$= \frac{133.33}{10}$$
$$= 13.33$$

Since $G_{1s} = 0$, $G_{2s} = 13.33$ percent and since $G_{2c} = 0$, $G_{1c} = 13.33$ percent. To check if the 20-mph design speed assumption was correct, the following equation must be shown to be correct:

$$\frac{AL_s}{200} + \frac{AL_c}{200} = 26.67$$

and, since substitution of the appropriate values into this equation show that the equality holds, the 20-mph design speed assumption was correct.

For stationing of curve points, it is clear that $PVI_s = 1 + 33.33$, $PVT_s = PVC_c = 2 + 66.67$, and $PVI_c = 3 + 33.33$. For elevations, $PVC_s = PVI_s = 100$ ft and $PVI_c = PVT_c = 126.67$ ft. Finally, the elevation of PVT_s and PVC_c can be computed as

$$100 + \frac{AL_s}{200} = 100 + \frac{13.33(266.67)}{200}$$
$$= 117.77 \text{ ft}$$

3.4 HORIZONTAL ALIGNMENT

The critical aspect of horizontal alignment is the horizontal curve that focuses on the design of the directional transition of the roadway in a horizontal plane. Stated differently, a horizontal curve provides a transition between two straight (or tangent) sections of roadway. A key concern in this directional transition is the ability of the vehicle to negotiate a horizontal curve (the provision of adequate drainage is also important, but it is not discussed in this book; see [AASHTO 1984]. As was the case with the straight-line vehicle performance characteristics discussed at length in Chapter 2, the highway engineer must design a horizontal alignment to accommodate a variety of vehicle cornering capabilities that range from nimble sports cars to ponderous trucks. A theoretical assessment of vehicle cornering at the level of detail given straightline performance in Chapter 2, is beyond the scope of this book; see [Campbell 1978; Wong 1978; and Brewer and Rice 1983]. Instead, vehicle cornering performance is viewed only at the practical design-oriented level, with equations simplified in a manner similar to the stopping-distance equation simplifications discussed in Section 2.9.4.

3.4.1 Vehicle Cornering

Figure 3.8 illustrates the forces acting on a vehicle during cornering. In this figure, α is the angle of incline, W is the weight of the vehicle (in pounds) with

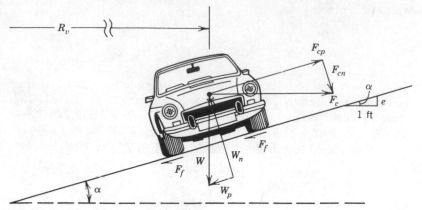

FIGURE 3.8
Vehicle cornering forces.

W_n and W_p being the weight normal and parallel to the roadway surface respectively, F_f is the side frictional force (centripetal, in pounds), F_c is the centrifugal force (lateral acceleration × mass, in pounds) with F_{cp} being the centrifugal force acting parallel to the roadway surface, F_{cn} is the centrifugal force acting normal to the roadway surface, and R_v is the radius defined to the vehicle's traveled path (in feet). Some basic horizontal curve relationships can be derived by summing forces parallel to the roadway surface,

$$W_p + F_f = F_{cp} \tag{3.22}$$

From basic physics this equation can be written as [with $F_f = f_s(W_n + F_{cn})$],

$$W \sin \alpha + f_s\left(W \cos \alpha + \frac{WV^2}{gR_v} \sin \alpha\right) = \frac{WV^2}{gR_v} \cos \alpha \tag{3.23}$$

where f_s is the coefficient of side friction (which is different from the coefficient of friction term, f, used in stopping-distance computations), g is the gravitational constant, and V is the vehicle speed (in feet per second). Dividing both sides of Eq. 3.23 by $W \cos \alpha$ gives

$$\tan \alpha + f_s = \frac{V^2}{gR_v}(1 - f_s \tan \alpha) \tag{3.24}$$

The term $\tan \alpha$ is referred to as the superelevation of the curve and is denoted e. In words, the superelevation is the number of vertical feet rise per 1 foot horizontal and has units of ft/ft (see Fig. 3.8). The term $f_s \tan \alpha$ in Eq. 3.24 is conservatively set equal to zero for practical applications due to the small values that f_s and α typically assume. With $e = \tan \alpha$, Eq. 3.24 can be arranged such that

$$R_v = \frac{V^2}{g(f_s + e)}$$

(3.25)

Example 3.6

A roadway is being designed for a speed of 70 mph. At one horizontal curve, it is known that the superelevation is 0.08 and the coefficient of side friction is 0.10. Determine the minimum radius of curve (measured to the traveled vehicle path) that will provide for safe vehicle operation.

Solution

The application of Eq. 3.25 gives

$$R_v = \frac{V^2}{g(f_s + e)} = \frac{(70 \times 1.47)^2}{32.2(0.08 + 0.1)}$$

$$= 1826.85 \text{ ft}$$

This value is the minimum radius, since radii larger than 1826.85 ft will generate centrifugal forces lower than those capable of being safely supported by the superelevation and the side frictional force.

In the actual design of a horizontal curve, the engineer must select appropriate values of e and f_s. The value selected for superelevation, e, is critical, since high rates of superelevation can cause vehicle steering problems at exits on horizontal curves and, in cold climates, ice on roadways can reduce f_s such that vehicles traveling less than design speed on an excessively superelevated curve could be forced inwardly off the curve by gravitational forces. AASHTO provides general guidelines for the selection of e and f_s for horizontal curve design, as shown in Table 3.4. The values presented in this table are grouped by four values of maximum e. The selection of any one of these four maximum e values is dependent on the type of road (e.g., higher maximum e's are permitted on freeways relative to arterials and local roads) and local design practice. Maximum f_s's are simply a function of the design speed.

TABLE 3.4
Maximum Degree of Curve and Minimum Radius
Determined for Limiting Values of
e and f_s

Design Speed (mph)	Maximum e	Maximum f_s
20	0.04	0.17
30	0.04	0.16
40	0.04	0.15
50	0.04	0.14
60	0.04	0.12
20	0.06	0.17
30	0.06	0.16
40	0.06	0.15
50	0.06	0.14
60	0.06	0.12
65	0.06	0.11
70	0.06	0.10
20	0.08	0.17
30	0.08	0.16
40	0.08	0.15
50	0.08	0.14
60	0.08	0.12
65	0.08	0.11
70	0.08	0.10
20	0.10	0.17
30	0.10	0.16
40	0.10	0.15
50	0.10	0.14
60	0.10	0.12
65	0.10	0.11
70	0.10	0.10

Source: American Association of State Highway and Transportation Officials, "A Policy on Geometric Design of Highways and Streets," Washington, D.C., 1984.

Note: In recognition of safety considerations, use of maximum $e = 0.04$ should be limited to urban conditions.

3.4.2 Horizontal Curve Fundamentals

In connecting straight (tangent) sections of roadway with a curve, several options are available. The most obvious of these is the simple curve, which is just a standard curve with a single, constant radius. Other options include compound curves, which consist of two or more simple curves in succession, and spiral curves, which are curves with a continuously changing radius. To illustrate the basic principles involved in horizontal curve design, this book will examine only simple curves. For information regarding compound and spiral curves, the reader is referred elsewhere [AASHTO 1984].

Figure 3.9 shows the basic elements of a simple horizontal curve. In this figure, R is the radius (measured to the centerline of the road), PC is the beginning point of the horizontal curve, T is the tangent length, PI is the tangent intersection, Δ is the central angle of the curve, PT is the ending point of the horizontal curve, M is the middle ordinate, E is the external distance, and L is the length of curve. Another important term is the degree of curve, which is defined as the angle subtended by a 100-ft arc along the horizontal curve. It is a measure of the sharpness of the curve and is frequently used instead of the radius in the actual construction of a horizontal curve. The degree of curve is directly related to the radius of the horizontal curve by

$$D = \frac{5729.6}{R} \tag{3.26}$$

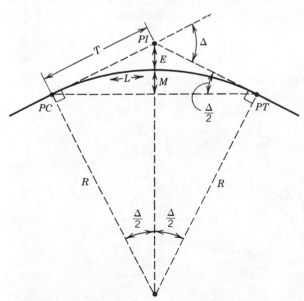

FIGURE 3.9
Elements of a simple horizontal curve.

A geometric and trigonometric analysis of Fig. 3.9 reveals the following relationships:

$$T = R \tan \frac{\Delta}{2} \tag{3.27}$$

$$E = R\left(\frac{1}{\cos(\Delta/2)} - 1\right) \tag{3.28}$$

$$M = R\left(1 - \cos \frac{\Delta}{2}\right) \tag{3.29}$$

$$L = \frac{100\Delta}{D} \tag{3.30}$$

It is important to note that horizontal curve stationing, curve length, and curve radius (R) are typically measured with respect to the centerline of the road. In contrast, the radius determined on the basis of vehicle forces (R_v in Eq. 3.25) is measured from the innermost vehicle path, which is assumed to be the midpoint of the innermost vehicle lane. Thus, to be truly correct, a slight correction for lane width is required when equating the R_v of Eq. 3.25 with the R in Eqs. 3.27, 3.28, and 3.29.

Example 3.7

A horizontal curve is designed with a 2000-ft radius. The curve has a tangent of 400 ft and the *PI* is at station 103 + 00. Determine the stationing of the *PT*.

Solution

Equation 3.27 is applied to determine the central angle, Δ.

$$T = R \tan \frac{\Delta}{2}$$

$$400 = 2000 \tan \frac{\Delta}{2}$$

$$\Delta = 22.62°$$

So, from Eqs. 3.30 and 3.26, the length of the curve is

$$L = \frac{100\Delta}{D}$$

$$= \frac{100\Delta}{\frac{5729.6}{R}}$$

$$= \frac{100(22.62)(2000)}{5729.6}$$

$$= 789.58 \text{ ft}$$

Given that the tangent is 400 ft,

$$\text{stationing } PC = 103 + 00 - 4 + 00$$
$$= 99 + 00$$

Since horizontal curve stationing is measured along the alignment of the road,

$$\text{stationing } PT = \text{stationing } PC + L$$
$$= 99 + 00 + 7 + 89.58$$
$$= 106 + 89.58$$

3.4.3 Stopping-Sight Distance and Horizontal Curve Design

As was the case for vertical curve design, adequate stopping-sight distance must also be provided in the design of horizontal curves. Sight-distance restrictions on horizontal curves occur when obstructions are present as shown in Fig. 3.10. Such obstructions are frequently encountered in highway design due to the cost of right-of-way acquisition and/or the cost of moving earthen materials (e.g., rock outcroppings). When such an obstruction exists, the stopping-sight distance is

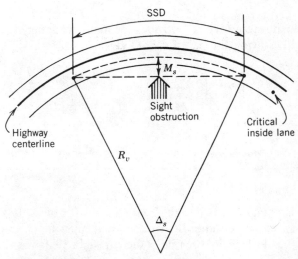

FIGURE 3.10
Stopping-sight distance considerations for horizontal curves.

measured along the horizontal curve from the center of the traveled lane (the assumed location of the driver's eyes). As shown in Fig. 3.10, for a specified stopping distance, some distance, M_s (the middle ordinate), must be visually cleared, so that the line of sight is such that sufficient stopping-sight distance is available.

Equations for computing SSD relationships for horizontal curves can be derived by first determining the central angle, Δ_s, for an arc equal to the required stopping-sight distance (see Fig. 3.10 and note that this is not the central angle of the horizontal curve whose arc is equal to L). Assuming that the length of the horizontal curve exceeds the required SSD, we have (as with Eq. 3.30),

$$\text{SSD} = \frac{100\Delta_s}{D} \tag{3.31}$$

Combining Eq. 3.31 with Eq. 3.26 gives

$$\Delta_s = \frac{57.296 \, \text{SSD}}{R_v} \tag{3.32}$$

where R_v is the radius to the vehicle's traveled path, which is also assumed to be the location of the driver's eyes for sight distance, and is again taken as the radius to the middle of the innermost lane, and Δ_s is the angle subtended by an arc equal to the SSD in length. Substituting Eq. 3.32 into the general equation for the middle ordinate of a simple horizontal curve (Eq. 3.29) gives

$$M_s = R_v \left[1 - \cos\left(\frac{28.65 \, \text{SSD}}{R_v} \right) \right] \tag{3.33}$$

where M_s is the middle ordinate necessary to provide adequate stopping-sight distance, as shown in Fig. 3.10. Solving Eq. 3.33 for SSD gives

$$\text{SSD} = \frac{R_v}{28.65} \left[\cos^{-1}\left(\frac{R_v - M_s}{R_v} \right) \right] \tag{3.34}$$

Example 3.8

A horizontal curve on a two-lane highway is designed with a 2000-ft radius, 12-ft lanes, and a 60-mph design speed. Determine the distance that must be cleared from the inside edge of the inside lane to provide sufficient sight distance for minimum SSD and desirable SSD.

Solution

The curve radius is typically given from the centerline of the roadway. Thus, for calculating the required distance to be cleared, 6 ft is subtracted from the radius to give the driver location (i.e., the middle of the traveled lane) for the critical inside lane. From Table 3.1, the minimum SSD is 525 ft, so applying Eq. 3.33 gives

$$M_s = R_v \left[1 - \cos \left(\frac{28.65 \text{ SSD}}{R_v} \right) \right]$$

$$= 1994 \left[1 - \cos \left(\frac{28.65 \, (525)}{1994} \right) \right]$$

$$= 17.26 \text{ ft}$$

Therefore, 17.26 ft must be cleared, as measured from the center of the inside lane, or 11.26 ft from the inside edge of the inside lane. Similarly, the desirable SSD is 650 ft (Table 3.1) and application of Eq. 3.33 gives $M_s = 26.43$ ft. Thus, for desirable SSD, 20.43 ft must be cleared from the inside edge of the inside lane.

Example 3.9

A two-lane highway (two 12-ft lanes) has a posted speed limit of 50 mph and, on one section, has both horizontal and vertical curves as shown in Fig. 3.11. A recent daytime accident (driver traveling eastbound and striking a stationary roadway object) resulted in a fatality and a law suit alleging that the 50-mph posted speed limit was an unsafe speed for the curves in question and a major cause of the accident. Evaluate and comment on the roadway design.

Solution

Begin with an assessment of the horizontal alignment. Two concerns must be considered: the adequacy of the curve radius and superelevation and the adequacy of the sight distance on the eastbound (inside) lane. For the curve radius,

Plan View (horizontal alignment)

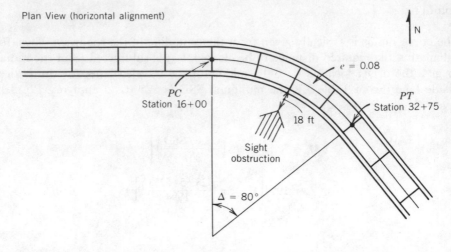

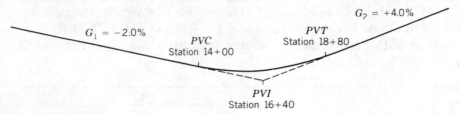

FIGURE 3.11
Horizontal and vertical alignment for Example 3.9.

note from the diagram that,

$$L = 32 + 75 - 16 + 00$$
$$= 16 + 75$$
$$= 1675 \text{ ft}$$

Using Eq. 3.30, we obtain

$$D = \frac{100\Delta}{L}$$

$$= \frac{100(80)}{1675}$$

$$= 4.78°$$

From Eq. 3.26, we get

$$R = \frac{5729.58}{4.78}$$

$$= 1198.65 \text{ ft}$$

Using the posted speed limit of 50 mph with $e = 0.08$, we find that Eq. 3.25 can be rearranged to give (with the vehicle traveling in the middle of the inside lane, $R_v = R -$ half the lane width or $R_v = 1198.65 - 6 = 1192.65$),

$$f_s = \frac{V^2}{gR_v} - e$$

$$= \frac{(50 \times 1.47)^2}{32.2(1192.65)} - 0.08$$

$$= 0.061$$

From Table 3.4, the maximum f_s for 50 mph is 0.14. Since 0.061 does not exceed 0.14, the radius and superelevation are sufficient for the 50-mph speed.

For the sight distance, the available M_s is 18 ft plus the 6-ft distance to the center of the eastbound (inside) lane, or 24 ft. Application of Eq. 3.34 gives (with $R_v = 1198.65 - 6 = 1192.65$, again subtracting half the lane width from the radius of the roadway's centerline)

$$\text{SSD} = \frac{R_v}{28.64}\left[\cos^{-1}\left(\frac{R_v - M_s}{R_v}\right)\right]$$

$$= \frac{1192.65}{28.65}\left[\cos^{-1}\left(\frac{1192.65 - 24}{1192.65}\right)\right]$$

$$= 479.3 \text{ ft}$$

From Table 3.1, SSD = 475 at 50 mph, so the 479.3 ft of SSD provided is sufficient. Turning to the sag vertical curve, the length of curve is

$$L = \text{station of PVT} - \text{station of PVI}$$
$$= 18 + 80 - 14 + 00$$
$$= 4 + 80 \quad \text{or} \quad 480 \text{ ft}$$

Also, from Fig. 3.11, $A = 6$. Using Eq. 3.10, we obtain

$$K = \frac{L}{A}$$

$$= \frac{480}{6}$$

$$= 80$$

For the 50-mph speed, Table 3.3 indicates a necessary K-value of 110 desirable and 90 minimum. Thus the K-value of 80 reveals that the vertical curve is underdesigned for the 50-mph speed. However, since the accident occurred in daylight and sight distances on sag vertical curves are governed by nighttime conditions, this design flaw is not likely to have contributed to the accident.

NOMENCLATURE
FOR
CHAPTER 3

A absolute value of the algebraic difference in grades (in percent)

D degree of curvature

E external distance

e rate of superelevation

F_f frictional side force

F_c centrifugal force

F_{cn} centrifugal force normal to the roadway surface

F_{cp} centrifugal force parallel to the roadway surface

f coefficient of stopping friction

f_s coefficient of side friction

G grade

G_1 initial roadway grade

G_2 final roadway grade

g gravitational constant

H height of vehicle headlights

H_1 height of driver's eye

H_2 height of roadway object

K horizontal distance required to affect a 1 percent change in slope

L length of curve

L_m minimum length of curve

M middle ordinate

M_s middle ordinate for stopping sight distance

PC	initial point of horizontal curve
PI	point of tangent intersection (horizontal curve)
PT	final point of horizontal curve
PVC	initial point of vertical curve
PVI	point of vertical intersection
PVT	final point of vertical curve
R	radius of curve measured to the roadway centerline
R_v	radius of curve measured to center of vehicle
S	sight distance
SSD	stopping-sight distance
T	tangent length
V	vehicle speed
V_1	initial vehicle speed
W	vehicle weight
W_n	vehicle weight normal to the roadway surface
W_p	vehicle weight parallel to the roadway surface
x	distance from the beginning of a vertical curve
Y	vertical curve offset
Y_f	end-of-curve offset
Y_m	vertical curve midcurve offset
α	angle of superelevation
β	angle of upward headlight beam
Δ	central angle
Δ_s	central angle subtended by the stopping-sight distance (SSD) arc

REFERENCES

1. American Association of State Highway and Transportation Officials, "A Policy on Geometric Design of Highways and Streets," Washington, D.C., 1984.

2. P. H. Wright, and R. J. Paquette, *Highway Engineering*, John Wiley & Sons, New York, 1987.

3. F. L. Mannering, "Computer Plotting of Highway Perspectives," unpublished Bachelor's Thesis, University of Saskatchewan, Saskatoon, Canada, 1976.

4. C. Campbell, "The Sports Car: Its Design and Performance," Robert Bently, Inc., Cambridge, Mass., 1978.

5. J. H. Wong, *Theory of Ground Vehicles*, John Wiley & Sons, New York, 1978.

6. H. K. Brewer, and R. S. Rice, "Tires: Stability and Control," *SAE Transactions*, Vol. 92, paper 830561, 1983.

PROBLEMS

3.1. A 1600-ft-long sag vertical curve (equal tangent) has a *PVC* at station 120 + 00 and elevation 1500 ft. The initial grade is −3.5 percent and the final grade is +6.5 percent. Determine the elevation and stationing of the low point, *PVI* and *PVT*.

3.2. A 500-ft-long, equal tangent crest vertical curve connects tangents that intersect at station 340 + 00 and elevation 1322 ft. The initial grade is +4.0 percent and the final grade is −2.5 percent. Determine the elevation and stationing of the high point, *PVC* and *PVT*.

3.3. A vertical curve crosses a 4-ft-diameter pipe at right angles. The pipe is located at station 110 + 85 and its centerline is at elevation 1091.60 ft. The *PVI* of the vertical curve is at station 110 + 00 and elevation 1098.4 ft. The curve is equal tangent, 600 ft long, and connects an initial grade of +1.20 percent and a final grade of −1.08 percent. Determine the depth, below the surface of the curve, of the top of the pipe.

3.4. Consider the curve described in Problem 3.3. Does this curve have adequate stopping-sight distance for a speed of 60 mph?

3.5. An equal tangent sag vertical curve is designed to provide desirable SSD. The *PVC* of the curve is at station 109 + 00 (elevation 950 ft), the *PVI* at station 110 + 92.5 (elevation 947.11 ft), and the low point is at station 110 + 65. Determine the design speed of the curve.

3.6. An equal tangent vertical curve was designed in 1988 (to 1984 AASHTO standards) for desirable SSD at a design speed of 70 mph to connect grades $G_1 = +1.0$ percent and $G_2 = -2.0$ percent. The curve is to be redesigned for a 70-mph design speed in the year 2020. Vehicle braking technology has advanced such that coefficients of stopping friction, f's, have increased by 40 percent relative to the 1984 standards given in Table 3.1, but, due to the higher percentage of older people in the driving population, design reaction times have increased by 20 percent. Also, vehicles have become smaller such that the driver's eye height is assumed to be 2.75 ft above the pavement and roadway objects are assumed to be 0.25 ft above the pavement. Compute the difference in design curve lengths for 1988 and 2020 designs.

3.7. A highway reconstruction project is being undertaken to reduce accident rates. The reconstruction involves a major realignment of the highway such that a 60-mph design speed is attained. At one point on the highway, an 800-ft equal tangent crest vertical curve exists. Measurements show that, at 3 + 52 stations from the *PVC*, the vertical curve offset is 3 ft. Assess the adequacy of the existing curve in light of the reconstruction design speed of 60 mph and, if the existing curve is inadequate, compute a satisfactory curve length. (Consider both minimum and desirable SSDs.)

3.8. Two level sections ($G = 0$) of an east–west highway, are to be connected. Currently, the two sections of highway are separated by a 4000-ft (horizontal distance), 2 percent grade. The westernmost section of highway is the higher of the two and is at elevation 100 ft. If the highway has a 60-mph design speed, determine, for the crest and sag vertical curves required, the stationing and elevation of the *PVC*s and *PVT*s given that the *PVC* of the crest vertical curve (on the westernmost level highway section) is at station 00 + 00 and elevation 100 ft. (assume desirable SSD and curve *PVI*s are at the intersection of $G = 0$ and the 2 percent grade, i.e., $A = 2$).

3.9. A horizontal curve is being designed through mountainous terrain for a four-lane road with 10-ft lanes. The central angle (Δ) is known to be 40°; the tangent distance is 510 ft; and the stationing of the tangent intersection (*PI*) is 2700 + 00. If the roadway surface has a coefficient of side friction of 0.082 and a superelevation of 0.09 ft/ft, determine the design speed and the stationing of the *PC* and *PT*.

3.10. A new interstate highway is being built with a design speed of 70 mph. For one of the horizontal curves, the radius (measured to the innermost vehicle path) is tentatively planned as 900 ft. What rate of superelevation is required at the design speed?

3.11. A developer is having a single-lane raceway constructed with a 100-mph design speed. A curve on the raceway has a radius of 1000 ft, a central angle of 30 degrees, and *PI* stationing at 1125 + 10. If the design coefficient of side friction is 0.2, determine the superelevation required at the design speed. Also, compute the degree of curve, length of curve, and stationing of the *PC* and *PT*.

3.12. A horizontal curve is being designed for a new two-lane highway (12-ft lanes). The *PI* is at station 250 + 50, design speed is 65 mph, and a maximum superelevation rate of 0.08 ft/ft is to be used. If the central angle of the curve is 35°, select a curve for the highway by computing the radius, degree of curve, and stationing of *PC* and *PT*.

3.13. You are asked to design a horizontal curve to connect tangents that intersect at $\Delta = 40°$ for a two-lane roadway (10-ft lanes) in Alaska. The design speed is 65 mph and, under the extreme Alaskan conditions, a maximum superelevation of 0.06 ft/ft is recommended. Give the radius, degree of curvature, and length of curve that you would recommend.

3.14. A freeway exit ramp has a single 12-ft lane and consists entirely of a curve with $\Delta = 90°$ and $L = 628$ ft. If the distance cleared from the centerline is 19.4 ft, what design speed was used? (Assume desirable SSD.)

3.15. Consider Example 3.9. What must be done to provide a safe 60-mph horizontal design speed using the same R with possibly new values of e, f_s, and/or M_s, with (a) desirable SSD and (b) minimum SSD?

3.16. For the curve in Problem 3.12, what distance must be cleared from the inside edge of the inside lane to provide adequate sight distance for minimum and desirable SSDs?

Chapter Four

Pavement Design

4.1 INTRODUCTION

The physical components of a highway system include the right of way, the pavement, shoulders, safety appurtenances, signs, signals, and markings. The pavement and shoulders represent the most costly items associated with highway construction and maintenance. In fact, the highway system in the United States is the most costly public works project undertaken by any society. Since the pavement and shoulder structures are the most expensive items to construct and maintain, it is important that the highway engineer has a basic understanding of pavement design principles.

In the United States, there are over 3 million miles of highways. It is surprising that many of these roads are not paved; instead, they are comprised of either stabilized soil or gravel. A stabilized material consists of aggregate material bound together with a cementing agent such as portland cement, lime fly ash, or asphaltic cement. Highways that carry larger amounts of traffic with heavy axle loads, require surfaces with asphalt concrete or portland cement concrete to provide for all weather operations. These types of pavements can cost upward of several million dollars per mile to construct. Some states such as Pennsylvania, Texas, Illinois, and California have large pavement construction and rehabilitation budgets of $800 + million per year; and, when coupled with their maintenance budgets, it is easy to see why the construction and maintenance of pavement infrastructure must be done in a cost-effective manner.

Pavements provide two basic functions. First, the pavement can help to guide the driver. The pavement and shoulder give the driver a visual perspective of the horizontal and vertical alignment of the travel path. Consequently, the driver is given information about the driving task and the steering of the vehicle. The second function of the pavement is to support the vehicle loads. It is this second function that will be discussed in this chapter.

4.2 TYPES OF PAVEMENT

In general, there are two types of pavement structures: a flexible pavement or a rigid pavement. There are, however, many variations of these pavement types. Some include soil cement and stabilized bases that have cemented aggregate. Composite pavements, which are made of both rigid and flexible layers and continuously reinforced, and post-tensioned pavements are other types of pavements. These other types of pavements usually require specialized designs, and therefore, they are not covered in this chapter.

As with any structure, the underlying soil must ultimately carry the load that is placed on it. It is a pavement's function to distribute the traffic loading stresses to the soil (subgrade) at a magnitude that will not shear or distort the soil.

Typical soil bearing capacities can be less than 50 pounds per square inch (psi) and in some cases as low as 2 to 3 psi. Also, when the soil is saturated with water, the bearing capacity can be very low; in these cases it is very important that the pavement distribute the tire loads to avoid pavement failure.

A typical automobile weighs approximately 2500 lb, with tire pressures of 35 psi. These loads are very small when compared to a typical semitrailer truck. Large trucks can carry loads up to 80,000 lb on five axles with tire pressures of 100 psi or higher. Truck loads such as these represent the standard type of loading that is used for pavement design. In this chapter, attention will be given to an accepted pavement design procedure that can be used to design pavement structures for high volume roadway facilities subjected to heavy truck traffic. The design of stabilized soil pavements and gravel-surfaced pavements can be found in other references [Yoder & Witczak, 1975].

4.2.1 Flexible Pavements

A flexible pavement is constructed with asphaltic cement and aggregates, as shown in Fig. 4.1. The pavement usually consists of several layers, as shown in the figure. The lower layer is called the subgrade, which is the soil itself. The upper 6 to 8 in. of subgrade is usually scarified and blended to provide a uniform material before it is compacted to maximum density. The next layer is the subbase, which usually consists of crushed aggregate (rock). This material has better engineering properties (higher modulus values) than the subgrade material, in terms of bearing capacity. Crushed aggregates from quarried rock sources and natural gravel deposits can be used for this layer. The next layer is called the base layer. This layer can be made of crushed aggregates, which are either unstabilized or stabilized with a cementing material. The cementing material can be portland cement, lime fly ash, or asphaltic cement. The base layer can be considered as the primary layer that distributes the traffic loads to the subbase and the subgrade.

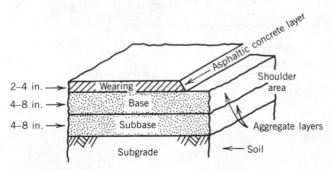

FIGURE 4.1
Typical flexible pavement cross section.

The top layer of a flexible pavement is called the wearing surface. It is usually made of asphaltic concrete, which is a mixture of asphalt cement and aggregates. The purpose of the wearing layer is to protect the base layer from wheel abrasion and to waterproof the entire pavement structure. It also provides a skid resistant surface that is important for safe vehicle stops. Typical thicknesses of the individual layers are shown in Fig. 4.1. These thicknesses will vary with the type of axle loading, available materials, and expected design life. The design life is the number of years the pavement is expected to provide adequate service before it must undergo major rehabilitation.

4.2.2 Rigid Pavements

A rigid pavement is constructed with portland cement concrete (PCC) and aggregates, as shown in Fig. 4.2. A rigid pavement can consist of several layers, as shown in the figure. The lower layer is called the subgrade and it consists of the in-situ soil. The soil is usually scarified, blended, and compacted to maximum density. The next layer is the base. This layer is optional, depending upon the engineering properties of the subgrade. If the subgrade soil is poor and erodable, then it is advisable that a base layer be used; however, if the soil has good engineering properties and drains well, a base layer can be eliminated. The top layer (wearing surface) is the portland cement concrete slab. Slab length varies from short slab spacing of 10 to 12 ft to a spacing of 40 ft or more. Transverse contraction joints are built into the pavement to control the cracking of the PCC concrete due to shrinkage of the concrete during the curing of the fresh concrete. Load transfer devices, such as dowel bars, can be placed in the joints to minimize deflections and reduce stresses near the edges of the slabs. Slab thicknesses for PCC highway pavements vary from 8 to 12 in., as shown in Fig. 4.2.

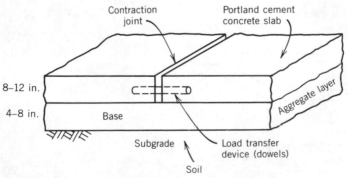

FIGURE 4.2
Typical rigid pavement cross section.

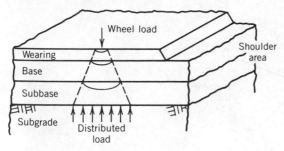

FIGURE 4.3
Distribution of load on a flexible pavement.

4.3 PAVEMENT SYSTEM DESIGN PRINCIPLES FOR FLEXIBLE PAVEMENTS

A primary function of the pavement system is to support the various traffic loads for the selected design life. As was stated earlier, the stresses caused by the vehicle loads are usually too large to be supported by the subgrade soil alone. Gravel surfaced pavements, on the other hand, can support heavy loads; however, the aggregate material can be abraded by tire action. Consequently, a pavement system must be designed to handle all of the adverse effects caused by traffic loads and the environment, including freeze thaw action.

In general, a pavement structure reduces and distributes the surface stresses (contact tire pressure) to an acceptable level at the subgrade. A flexible pavement reduces the stresses by the strength of the individual layers. The traffic loads are transmitted to the subgrade by aggregate-to-aggregate particle contact. Confining pressures (lateral forces due to material weight) in the subbase and base layers

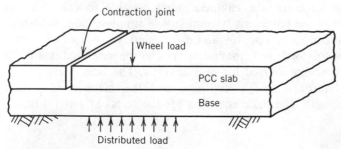

FIGURE 4.4
Beam action of a rigid pavement.

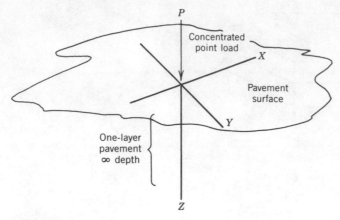

FIGURE 4.5
Point load on a one-layer pavement.

increase the bearing strength of these materials. A cone of distributed loads reduces and spreads the stresses to the subgrade, as shown in Fig. 4.3.

A rigid pavement, on the other hand, distributes the wheel loads by the beam action of the portland cement concrete (PCC) slab. The PCC slab is made of a material that has a high modulus of elasticity (4 to 5 million psi) compared to the other pavement materials. The modulus of elasticity, known as Young's modulus, is the ratio of stress to strain as a load is applied to the material. Therefore, the beam action (see Fig. 4.4) distributes the loads over a large area of the pavement, thus reducing large surface stresses to a level that is acceptable to the subgrade soil.

4.3.1 Calculation of Flexible Pavement Stresses and Deflections

To design a pavement structure, the engineer must be able to calculate the stresses and deflections in the pavement system. In the simplest case, the wheel load can be assumed to consist of a point load on a single-layer system, as shown in Fig. 4.5. This type of load and configuration can be analyzed with the Boussinesq solutions that were derived for soils analysis. The Boussinesq theory assumes that the pavement is one layer thick and the material is elastic, homogeneous, and isotropic. The basic equation for the stress at a point in the system is

$$\sigma_Z = K\frac{P}{z^2} \tag{4.1}$$

where P is the wheel load in pounds, z is the depth of the point in question in inches, and K is a variable defined as

$$K = \frac{3}{2\pi} \frac{1}{\left[1 + (r/z)^2\right]^{5/2}} \qquad (4.2)$$

where r is the radial distance in inches from the centerline of the point load to the point in question.

The Boussinesq theory is useful to begin the study of pavement stress calculations; however, it is not very representative of a pavement system loading and configuration, since it applies to a point load on one layer. A more realistic approach is to expand the point load to an elliptical area that represents a tire footprint. The tire footprint can be defined by an equivalent circular area with a radius calculated by

$$a = \sqrt{\frac{P}{p\pi}} \qquad (4.3)$$

where P is the tire load in pounds, p is the tire pressure in psi, and a is the equivalent load radius in inches. The integration of the load from a point to a circular area can be used to determine the stresses and deflections in a one-layer pavement system.

Several researchers have developed influence charts, graphical solutions, and equations for the calculation of stresses and deflections, which take into account a circular load. Ahlvin and Ulery provided solutions for the evaluation of stresses, strains, and deflections at any point in a homogeneous half-space [Ahlvin and Ulery 1962]. Their work made it easier to analyze a pavement system that was more complicated than the Boussinesq example. The one-layer equations by Ahlvin and Ulery can be used for a material with any Poisson ratio. The Poisson ratio describes the change in width to length when a load is applied along the vertical axis. The Ahlvin and Ulery equation for the calculation of vertical stress σ_z is

$$\sigma_z = p[A + B] \qquad (4.4)$$

where p is the pressure due to the load in psi, and A and B are function values that depend on z/a and r/a, the depth in radii and offset distance in radii, respectively. The variables z, a, and r are as defined for the Boussinesq and circular load equations. The equation for radial horizontal stress, σ_r, which causes pavement cracking is

$$\sigma_r = p[2\mu A + C + (1 - 2\mu)F] \qquad (4.5)$$

and the equation for deflection, Δ_z, is

$$\Delta_z = \frac{p(1 + \mu)a}{E_1}[zA + (1 - \mu)H] \qquad (4.6)$$

where E is the modulus of elasticity, as previously defined, and F and H are function values. The function values A, B, C, F, and H are presented in Table 4.1.

TABLE 4.1
One-Layer Elastic Function Values

Function A

Depth (z) in Radii	Offset (r) in Radii																
	0	0.2	0.4	0.6	0.8	1	1.2	1.5	2	3	4	5	6	8	10	12	14
0	1.0	1.0	1.0	1.0	1.0	0.5	0	0	0	0	0	0	0	0	0	0	0
0.1	0.90050	0.89748	0.88679	0.86126	0.78797	0.43015	0.09645	0.02787	0.00856	0.00211	0.00084	0.00042					
0.2	0.80388	0.79824	0.77884	0.73483	0.63014	0.38269	0.15433	0.05251	0.01680	0.00419	0.00167	0.00083	0.00048	0.0020			
0.3	0.71265	0.70518	0.68316	0.62690	0.52081	0.34375	0.17964	0.07199	0.02440	0.00622	0.00250						
0.4	0.62861	0.62015	0.59241	0.53767	0.44329	0.31048	0.18709	0.08593	0.03118								
0.5	0.55279	0.54403	0.51622	0.46448	0.38390	0.28156	0.18556	0.09499	0.03701	0.01013	0.00407	0.00209	0.00118	0.0053	0.00025	0.00014	0.00009
0.6	0.48550	0.47691	0.45078	0.40427	0.33676	0.25588	0.17952	0.10010									
0.7	0.42654	0.41874	0.39491	0.35428	0.29833	0.21727	0.17124	0.10228	0.04558								
0.8	0.37531	0.36832	0.34729	0.31243	0.26581	0.21297	0.16206	0.10236									
0.9	0.33104	0.32492	0.30669	0.27707	0.23832	0.19488	0.15253	0.10094									
1	0.29289	0.28763	0.27005	0.24697	0.21468	0.17868	0.14329	0.09849	0.05185	0.01742	0.00761	0.00393	0.00226	0.00097	0.00050	0.00029	0.00018
1.2	0.23178	0.22795	0.21662	0.19890	0.17626	0.15101	0.12570	0.09192	0.05260	0.01935	0.00871	0.00459	0.00269	0.00115	0.00073	0.00043	0.00027
1.5	0.16795	0.16552	0.15877	0.14804	0.13436	0.11892	0.10296	0.08048	0.05116	0.02142	0.01013	0.00548	0.00325	0.00141	0.00094	0.00056	0.00036
2	0.10557	0.10453	0.10140	0.09647	0.09011	0.08269	0.07471	0.06275	0.04496	0.02221	0.01160	0.00659	0.00399	0.00180	0.00115	0.00068	0.00043
2.5	0.07152	0.07098	0.06947	0.06698	0.06373	0.05974	0.05555	0.04880	0.03787	0.02143	0.01221	0.00732	0.00463	0.00214	0.00132	0.00079	0.00051
3	0.05132	0.05101	0.05022	0.04886	0.04707	0.04487	0.04241	0.03839	0.03150	0.01980	0.01220	0.00770	0.00505	0.00242	0.00160	0.00099	0.00065
4	0.02986	0.02976	0.02907	0.02832	0.02802	0.02749	0.02651	0.02490	0.02193	0.01592	0.01109	0.00768	0.00536	0.00282	0.00179	0.00113	0.00075
5	0.01942	0.01938				0.01835			0.01573	0.01249	0.00949	0.00708	0.00527	0.00298	0.00188	0.00124	0.00084
6	0.01361					0.01307			0.01168	0.00983	0.00795	0.00628	0.00492	0.00299	0.00193	0.00130	0.00091
7	0.01005					0.00976			0.00894	0.00784	0.00661	0.00548	0.00445	0.00291	0.00189	0.00134	0.00094
8	0.00772					0.00755			0.00703	0.00635	0.00554	0.00472	0.00398	0.00276	0.00184	0.00133	0.00096
9	0.00612					0.00600			0.00566	0.00520	0.00466	0.00409	0.00353	0.00256			
10								0.00477	0.00465	0.00438	0.00397	0.00352	0.00326	0.00241			

TABLE 4.1 *continued*

Function B

Depth (z) in Radii	Offset (r) in Radii																
Radii	0	0.2	0.4	0.6	0.8	1	1.2	1.5	2	3	4	5	6	8	10	12	14
0	0	0	0	0	0	0	0	0	0	0	0	0	0	0	0	0	0
0.1	0.09852	0.10140	0.11138	0.13424	0.18796	0.05388	−0.07899	−0.02672	−0.00845	−0.00210	−0.00084	0	0	0	0	0	0
0.2	0.18857	0.19306	0.20772	0.23524	0.25983	0.08513	−0.07759	−0.04448	−0.01593	−0.00412	−0.00166	−0.00042	0	0	0	0	0
0.3	0.28362	0.26787	0.28018	0.29483	0.27257	0.10757	−0.04316	−0.04999	−0.02166	−0.00599	−0.00245	−0.00083	−0.00024	−0.00010			
0.4	0.32016	0.32259	0.32748	0.32273	0.26925	0.12404	−0.00766	−0.04535	−0.02522								
0.5	0.35777	0.35752	0.35323	0.33106	0.26236	0.13591	0.02165	−0.03455	−0.02651	−0.00991	−0.00388	−0.00199	−0.00116	−0.00049	−0.0025	−0.0014	−0.00009
0.6	0.37831	0.37531	0.36308	0.32822	0.25411	0.14440	0.04457	−0.02101									
0.7	0.38487	0.37962	0.36072	0.31929	0.24638	0.14986	0.06209	−0.00702	−0.02329								
0.8	0.38091	0.37408	0.35133	0.30699	0.23779	0.15292	0.07530	0.00614									
0.9	0.36962	0.36275	0.33734	0.29299	0.22891	0.15404	0.08507	0.01795									
1	0.35355	0.34553	0.32075	0.27819	0.21978	0.15355	0.09210	0.02814	−0.01005	−0.01115	−0.00608	−0.00344	−0.00210	−0.00092	−0.00048	−0.00028	−0.00018
1.2	0.31485	0.30730	0.28481	0.24836	0.20113	0.14915	0.10002	0.04378	0.00023	−0.00995	−0.00632	−0.00378	−0.00236	−0.00107			
1.5	0.25602	0.25025	0.23338	0.20694	0.17368	0.13732	0.10193	0.05745	0.01385	−0.00669	−0.00600	−0.00401	−0.00265	−0.00126	−0.00068	−0.00040	−0.00026
2	0.17889	0.18144	0.16644	0.15198	0.13375	0.11331	0.09254	0.06371	0.02836	0.00028	−0.00410	−0.00371	−0.00278	−0.00148	−0.00084	−0.00050	−0.00033
2.5	0.12807	0.12633	0.12126	0.11327	0.10298	0.09130	0.07869	0.06022	0.03429	0.00661	−0.00130	−0.00271	−0.00250	−0.00156	−0.00094	−0.00059	−0.00039
3	0.09487	0.09394	0.09099	0.08635	0.08033	0.07325	0.06551	0.05354	0.03511	0.01112	0.00157	−0.00134	−0.00192	−0.00151	−0.00099	−0.00065	−0.00046
4	0.05707	0.05666	0.05562	0.05383	0.05145	0.04773	0.04532	0.03995	0.03066	0.01515	0.00595	0.00155	−0.00109	−0.00109	−0.00094	−0.00068	−0.00050
5	0.03772	0.03760				0.03384			0.02474	0.01522	0.00810	0.00371	−0.00029	−0.00043	−0.00070	−0.00068	−0.00049
6	0.02666					0.02468			0.01968	0.01380	0.00867	0.00496	0.00132	0.00028	−0.00037	−0.00047	−0.00045
7	0.01980					0.01868			0.01577	0.01204	0.00842	0.00547	0.00254	0.00093	−0.00002	−0.00029	0.00037
8	0.01526					0.01459			0.01279	0.01034	0.00779	0.00554	0.00332	0.00141	0.00035	−0.00008	−0.00025
9	0.01212					0.01170			0.01054	0.00888	0.00705	0.00533	0.00386	0.00178	0.00066	0.00012	−0.00012
10								0.00924	0.00879	0.00764	0.00631	0.00501	0.00382	0.00199			

TABLE 4.1 continued

Function C

Offset (r) in Radii

Depth (z) in Radii \ Radii	0	0.2	0.4	0.6	0.8	1	1.2	1.5	2	3	4	5	6	8	10	12	14
0	0	0	0	0	0	0	0	0	0	0	0	0	0	0	0	0	0
0.1	−0.04926	−0.05142	−0.05903	−0.07708	−0.12108	0.02247	0.12007	0.04475	0.01536	0.00403	0.00164	0.00082	0	0	0	0	0
0.2	−0.09429	−0.09755	−0.10872	−0.12977	−0.14552	0.02419	0.14896	0.07892	0.02951	0.00796	0.00325	0.00164	0.00094	0.00039	0	0	0
0.3	−0.13181	−0.13484	−0.14415	−0.15023	−0.12990	0.01988	0.13394	0.09816	0.04148	0.01169	0.00483						
0.4	−0.16008	−0.16188	−0.16519	−0.15985	−0.11168	0.01292	0.11014	0.10422	0.05067								
0.5	−0.17889	−0.17835	−0.17497	−0.15625	−0.09833	0.00483	0.08730	0.10125	0.05690	0.01824	0.00778	0.00399	0.00231	0.00098	0.00050	0.00029	0.00018
0.6	−0.18915	−0.18663	−0.17336	−0.14934	−0.08967	−0.00304	0.06731	0.09313									
0.7	−0.19244	−0.18831	−0.17393	−0.14147	−0.08409	−0.01061	0.05028	0.08253	0.06129								
0.8	−0.19046	−0.18481	−0.16784	−0.13393	−0.08066	−0.01744	0.03582	0.07114									
0.9	−0.18481	−0.17841	−0.16024	−0.12664	−0.07828	−0.02337	0.02359	0.05993									
1	−0.17678	−0.17050	−0.15188	−0.11995	−0.07634	−0.02843	0.01331	0.04939	0.05429	0.02726	0.01333	0.00726	0.00433	0.00188	0.00098	0.00057	0.00036
1.2	−0.15742	−0.15117	−0.13467	−0.10763	−0.07289	−0.03575	−0.00245	0.03107	0.04522	0.02791	0.01467	0.00824	0.00501	0.00221			
1.5	−0.12801	−0.12277	−0.11101	−0.09145	−0.06711	−0.04124	−0.01702	0.01088	0.03154	0.02652	0.01570	0.00933	0.00585	0.00266	0.00141	0.00083	0.00039
2	−0.08944	−0.08491	−0.07976	−0.06925	−0.05560	−0.04144	−0.02687	−0.00782	0.01267	0.02070	0.01527	0.01013	0.00321	0.00327	0.00179	0.00107	0.00069
2.5	−0.06403	−0.06068	−0.05839	−0.05259	−0.04522	−0.03605	−0.02800	−0.01536	0.00103	0.0134	0.00987	0.00707	0.00569	0.00209	0.00128	0.00083	
3	−0.04744	−0.04560	−0.04339	−0.04089	−0.03642	−0.03130	−0.02587	−0.01748	−0.00528	0.00792	0.01030	0.00888	0.00689	0.00392	0.00232	0.00145	0.00096
4	−0.02854	−0.02737	−0.02562	−0.02585	−0.02421	−0.02112	−0.01964	−0.01586	−0.00956	0.00038	0.00492	0.00602	0.00561	0.00389	0.00254	0.00168	0.00115
5	−0.01886	−0.01810				−0.01568			−0.00939	−0.00293	−0.00128	0.00329	0.00391	0.00341	0.00250	0.00177	0.00127
6	−0.01333					−0.01118			−0.00819	−0.00405	−0.00079	0.00129	0.00234	0.00272	0.00227	0.00173	0.00130
7	−0.00990					−0.00902			−0.00678	−0.00417	−0.00180	−0.00004	0.00113	0.00200	0.00193	0.00161	0.00128
8	−0.00763					−0.00699			−0.00552	−0.00393	−0.00225	−0.00077	0.00029	0.00134	0.00157	0.00143	0.00120
9	−0.00607					−0.00423			−0.00452	−0.00353	−0.00235	−0.00118	−0.00027	0.00082	0.00124	0.00122	0.00110
10						−0.00381			−0.00373	−0.00314	−0.00233	−0.00137	−0.0063	0.00040			

TABLE 4.1 *continued*

Function F

Depth (z) in Radii (rows) × Offset (r) in Radii (columns)

Depth (z) in Radii	0	0.2	0.4	0.6	0.8	1	1.2	1.5	2	3	4	5	6	8	10	12	14
0	0.5	0.5	0.5	0.5	0.5	0	−0.34722	−0.22222	−0.12500	−0.05556	−0.03125	−0.02000	−0.01389	−0.00781	−0.00500	−0.00347	−0.00255
0.1	0.45025	0.44794	0.43981	0.41954	0.35789	0.03817	−0.20800	−0.17612	−0.10950	−0.05151	−0.02961	−0.01917					
0.2	0.40194	0.39781	0.38294	0.34823	0.26215	0.05466	−0.11165	−0.13381	−0.09441	−0.04750	−0.02798	−0.01835	−0.01295	−0.00742			
0.3	0.35633	0.35094	0.34508	0.29016	0.20503	0.06372	−0.05346	−0.09768	−0.08010	−0.04356	−0.02636						
0.4	0.31431	0.30801	0.28681	0.24469	0.17086	0.06848	−0.01818	−0.06835	−0.06684								
0.5	0.27639	0.26997	0.24890	0.20937	0.14752	0.07037	0.00388	−0.04529	−0.05479	−0.03595	−0.02320	−0.01590	−0.01154	−0.00681	−0.00450	−0.00318	−0.00237
0.6	0.24275	0.23444	0.21667	0.18138	0.13042	0.07068	0.01797	−0.02749									
0.7	0.21327	0.20762	0.18956	0.15903	0.11740	0.06963	0.02704	−0.01392	−0.03469								
0.8	0.18765	0.18287	0.16679	0.14053	0.10604	0.06774	0.03277	−0.00365									
0.9	0.16552	0.16158	0.14747	0.12528	0.09664	0.06533	0.03619	0.00408									
1	0.14645	0.14280	0.12395	0.11225	0.08850	0.06256	0.03819	0.00984	−0.01367	−0.01994	−0.01591	−0.01209	−0.00931	−0.00587	−0.00400	−0.00289	−0.00219
1.2	0.11589	0.11360	0.10460	0.09449	0.07486	0.05670	0.03913	0.01716	−0.00452	−0.01491	−0.01337	−0.01068	−0.00844	−0.00550	−0.00353	−0.00261	−0.00201
1.5	0.08398	0.08196	0.07719	0.06918	0.05919	0.04804	0.03686	0.02177	0.00413	−0.00879	−0.00995	−0.00870	−0.00723	−0.00495	−0.00307	−0.00233	−0.00183
2	0.05279	0.05348	0.04994	0.04614	0.04162	0.03593	0.03029	0.02197	0.01043	−0.00189	−0.00546	−0.00589	−0.00544	−0.00410	−0.00263	−0.00208	−0.00166
2.5	0.03576	0.03673	0.03459	0.03263	0.03014	0.02762	0.02406	0.01927	0.01188	0.00198	−0.00226	−0.00364	−0.00386	−0.00332	−0.00223	−0.00183	−0.00150
3	0.02566	0.02586	0.02255	2595	0.02263	0.02097	0.01911	0.01623	0.01144	0.00396	−0.00010	−0.00192	−0.00258	−0.00263	−0.00153	−0.00137	−0.00120
4	0.01493	0.01536	0.01412	0.01259	0.01386	0.01331	0.01256	0.01134	0.00912	0.00508	0.00209	0.00026	−0.00076	−0.00148	−0.00096	−0.00099	−0.00093
5	0.00971	0.01011				0.00905			0.00700	0.00475	0.00277	0.00129	0.00031	−0.00066	−0.00053	−0.00066	−0.00070
6	0.00680					0.00675			0.00538	0.00409	0.00278	0.00170	0.00088	−0.00010	−0.00020	−0.00041	−0.00049
7	0.00503					0.00483			0.00428	0.00346	0.00258	0.00178	0.00114	0.00027	0.00003	−0.00020	−0.00033
8	0.00386					0.00380			0.00350	0.00291	0.00229	0.00174	0.00125	0.00048	0.00020	−0.00005	−0.00019
9	0.00306					0.00374			0.00291	0.00247	0.00203	0.00163	0.00124	0.00062			
10								0.00267	0.00246	0.00213	0.00176	0.00149	0.00126	0.00070			

TABLE 4.1 continued

Function H

Depth (z) in Radii	Offset (r) in Radii																
	0	0.2	0.4	0.6	0.8	1	1.2	1.5	2	3	4	5	6	8	10	12	14
0	2.0	1.97987	1.91751	1.80575	1.62553	1.27319	0.93676	0.71185	0.51671	0.33815	0.25200	0.20045	0.16626	0.12576	0.09918	0.08346	0.07023
0.1	1.80998	1.79018	1.72886	1.61961	1.44711	1.18107	0.92670	0.70888	0.51627	0.33794	0.25184	0.20081	0.16688	0.12512	0.09996	0.08295	0.07123
0.2	1.63961	1.62068	1.56242	1.46001	1.30614	1.09996	0.90098	0.70074	0.51382	0.33726	0.25162	0.20072	0.16668	0.12493	0.09952	0.08292	0.07104
0.3	1.48806	1.47044	1.40979	1.32442	1.19210	1.02740	0.86726	0.68823	0.50966	0.33638	0.25124	0.19982	0.16516	0.12394	0.09876	0.08270	0.07064
0.4	1.35407	1.33802	1.28963	1.20822	1.09555	0.96202	0.83042	0.67238	0.50412	0.33293	0.24996	0.19673	0.16369	0.12350	0.09792	0.08196	0.07026
0.5	1.23607	1.22176	1.17894	1.10830	1.01312	0.90298	0.79308	0.65429	0.49728								
0.6	1.13238	1.11998	1.08350	1.02154	0.94120	0.84917	0.75653	0.63469									
0.7	1.04131	1.03037	0.99794	0.91049	0.87742	0.80030	0.72143	0.61442	0.48061								
0.8	0.96125	0.95175	0.92386	0.87928	0.82136	0.75571	0.68809	0.59398									
0.9	0.89072	0.88251	0.85856	0.82616	0.77950	0.71495	0.65677	0.57361									
1	0.82843	0.85005	0.80465	0.76809	0.72587	0.67769	0.62701	0.55364	0.45122	0.31877	0.24386	0.19520	0.16199	0.12281	0.09700	0.08115	0.06980
1.2	0.72410	0.71882	0.70370	0.67937	0.64814	0.61187	0.57329	0.51552	0.43013	0.31162	0.24070	0.19053	0.15846	0.12124	0.09558	0.08061	0.06897
1.5	0.60555	0.60233	0.57246	0.57633	0.55559	0.53138	0.50496	0.46379	0.39872	0.29945	0.23495	0.18618	0.15395	0.11928	0.09300	0.07864	0.06848
2	0.47214	0.47022	0.44512	0.45656	0.44502	0.43202	0.41702	0.39242	0.35054	0.27740	0.22418	0.17898	0.14919	0.11694	0.08915	0.07675	0.06695
2.5	0.38518	0.38403	0.38098	0.37608	0.36940	0.36155	0.35243	0.33698	0.30913	0.25550	0.21208	0.17154	0.13864	0.11172	0.08562	0.07452	0.06522
3	0.32457	0.32403	0.32184	0.31887	0.31464	0.30969	0.30381	0.29364	0.27453	0.23487	0.19977	0.15596	0.12785	0.10585	0.08197	0.07210	0.06377
4	0.24620	0.24588	0.24820	0.25128	0.24168	0.23932	0.23668	0.23164	0.22188	0.19908	0.17640	0.14130	0.11778	0.09990	0.07800	0.06928	0.06200
5	0.19805	0.19785				0.19455			0.18450	0.17080	0.15575	0.12792	0.10843	0.09387	0.07407	0.06678	0.05976
6	0.16554					0.16326			0.15750	0.14868	0.13842	0.11620	0.09976	0.08848			
7	0.14217					0.14077			0.13699	0.13097	0.12404	0.10600	0.09234	0.08298			
8	0.12448					0.12352			0.12112	0.11680	0.11176	0.09702	0.08300	0.07710			
9	0.11079					0.10989			0.10854	0.10548	0.10161	0.08980					
10								0.09900	0.09820	0.09510	0.09290						

Example 4.1

A tire with 100 psi air pressure distributes a load over an area with a circular contact radius, a, of 5 in. The pavement was constructed with a material that has a modulus of elasticity of 50,000 psi and a Poisson ratio of 0.45. Calculate the radial horizontal stress and deflection at a point on the pavement surface under the center of the load. Also, calculate the radial horizontal stress and deflection at a point at a depth of 20 in. and a radial distance of 10 in.

Solution

With $z = 0$ in. and $r = 0$ in.

$$\frac{z}{a} = \frac{0}{5} = 0 \quad \text{and} \quad \frac{r}{a} = \frac{0}{5} = 0$$

From Table 4.1 for the above values

$$A = 1.0, \quad B = 0, \quad C = 0, \quad F = 0.5, \quad \text{and } H = 2.0$$

The radial horizontal stress is calculated from Eq. 4.5:

$$\sigma_r = p[2\mu A + C + (1 - 2\mu) F]$$
$$\sigma_r = 100[2(0.45)(1.0) + 0 + (1 - 2(0.45))(0.5)]$$
$$\sigma_r = 95.0 \text{ psi}$$

The deflection is calculated from Eq. 4.6.

$$\Delta_z = \frac{p(1 + \mu) a}{E_1} [zA + (1 - \mu) H]$$

$$= \frac{100(1 + 0.45)5}{50,000} [0(1.0) + (1 - 0.45)2.0]$$

$$= 0.016 \text{ in.}$$

where $z = 20.0$ in. and $r = 10.0$ in.
$z/a = 20/5 = 4$ and $r/a = 10/5 = 2$
$A = 0.02193$, $B = 0.03066$, $C = 0.00956$, $F = 0.00912$, and
$H = 0.22188$

The radial horizontal stress is

$$\sigma_r = 100[2(0.45)(0.02193) + (-0.00956) + (1 - 2(0.45))(0.00912)]$$
$$= 1.109 \text{ psi}$$

The deflection is

$$\Delta_z = \frac{100(1 + 0.45)^5}{50,000}\left[\frac{20}{5}(0.02193) + (1 - 0.045)(0.22188)\right]$$

$$= 0.008 \text{ in.}$$

The Boussinesq theory and the Ahlvin and Ulery equations can be used to calculate the stresses and deflections in a simple pavement system. The utility of the theory is that it serves as a basis for more complex pavement analysis. Today with the availability of the computer (both main frame and personal) there have been significant advances in mechanistic pavement analysis methodologies. Pavement systms are represented as multilayer systems that are homogenous and isotropic with linear responses. A linear response means that the material will return to its original shape after the load is removed. The programs allow the engineer to input various axle loads, tire pressures, and material properties. Program outputs consist of stresses, strains, and deflections. Some of the programs which are now in common use are: BISAR, CHEVRON, and ELSYM5. The ELSYM5 program is available on floppy disk from the Federal Highway Administration. The use of programs like ELSYM5 can make the task of pavement analysis much easier for the highway engineer.

4.4 FLEXIBLE PAVEMENT DESIGN PROCEDURE

A pavement system is required to perform several functions including vehicle guidance and support. In this section we will discuss the second function: providing support for the wheel loads. Beside distribution of the wheel loads, the pavement must also prevent moisture from entering the pavement system. As discussed, the intrusion of moisture in the system usually will weaken the subgrade and subbase. Material properties and the effect of moisture are topics beyond the scope of this book; see [Yoder and Witczak 1975]. It is the intent of the following sections to discuss a pavement design procedure that is used to determine layer thicknesses, rather than discuss individual materials properties.

4.4.1 The AASHTO Flexible Pavement Design Procedure

There are several accepted flexible pavement design procedures available to the highway engineer. Some examples are the Asphalt Institute Method and the Shell Procedure. Most of the procedures have been field-verified and used by highway agencies for several years. The selection of one procedure over another is usually based on highway agency experience and satisfaction with design results.

A widely accepted flexible pavement design procedure is presented in the "AASHTO Guide for Design of Pavement Structures," which is published by the

American Association of State Highway and Transportation Officials (AASHTO). The procedure was first published in 1972, with the latest revisions in 1986. Test data that were used for the development of the design procedure were collected at the AASHO (AASHO was the prior name) Road Test in Illinois from 1958 to 1960.

A pavement system can be subjected to many detrimental effects. It is subjected to various traffic loads that can produce fatigue failures. The fatigue failure (cracking) is the result of repeated loading due to traffic passing over the pavement. The pavement is also placed in an uncontrolled environment that produces temperature extremes and moisture variations. The combination of the environment, traffic loads, material variations, and construction variations requires a complex set of design procedures to incorporate all variables. The procedure, however, cannot be too complex or nobody will be able to use it. The AASHTO pavement design procedure meets most of the demands placed on a flexible pavement design procedure. It incorporates environment, load, and materials in a methodology that is not complicated to use. The procedure is empirical; however, the procedure has been widely accepted throughout the United States and around the world.

Before the details of the design procedure are discussed, it is important to understand the fundamental principles of the AASHTO design procedure. The supporting philosophy of the procedure is applicable to both flexible and rigid pavements (rigid pavement design is discussed in Section 4.5).

Serviceability Concept
Prior to the AASHO Road Test there was no real consensus as to the definition of pavement failure. In the eyes of an engineer, pavement failure occurred whenever cracking, rutting, or other surface distresses became visible. On the other hand, the motoring public usually associated pavement failure with a poor ride. The pavement engineers at the AASHO Road Test were faced with the task of combining the two failure definitions so that a single design procedure could be used to satisfy both critics. The Pavement Serviceability-Performance Concept was developed by Carey and Irick [1962] to handle the question concerning pavement failure. Carey and Irick proposed that a pavement, in general, will have a performance history. A pavement usually begins its service life in excellent condition; but, as traffic loading is applied, and with the interaction of the environment, the condition of the pavement will deteriorate until the pavement reaches an unserviceable level. The performance curve is the historical record of the performance of the pavement. Pavement performance, at any point in time, is known as the present serviceability index or PSI. Examples of pavement performance (or PSI trends) are shown in Fig. 4.6.

At any time the present serviceability index of the pavement can be measured. It is usually measured by a panel of raters that drive over the pavement section and rate the pavement performance on a scale of one to five with five being the smoothest ride. The accumulation of traffic loads will cause the pavement to deteriorate; as expected, the serviceability rating will drop. At some point, a

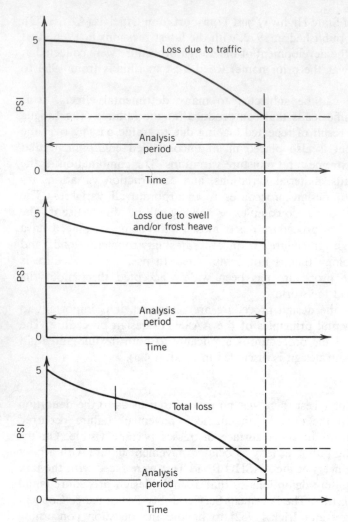

FIGURE 4.6
Pavement performance trends. Redrawn from "AASHTO Guide for Design of Pavement Structure," Washington, D.C., The American Association of State Highway and Transportation Officials, copyright 1986. Used by permission.

terminal serviceability index (TSI) is reached. At this point, most raters feel that the pavement can no longer perform in a serviceable manner.

It should be obvious that it would be a tremendous job to have panels of raters evaluate all of the nation's highways on a continuous basis. Instead, correlations have been made between panel opinions and measured variables such as pavement roughness, rutting, and cracking. Consequently, mechanical devices are now used to determine PSI rather than with a panel of raters.

It has been found that new pavements usually have an initial PSI rating of approximately 4.2 to 4.5. As traffic is applied to the pavement and with the climate interaction with the pavement, the PSI will continue to drop to an unacceptable level. This level or terminal serviceability index, TSI, is selected on the basis of the highway function. A highway facility such as an interstate highway will usually have a TSI of 3.0 while a local road can have a TSI of 2.0.

Flexible Pavement Design Equation

At the conclusion of the AASHO Road Test a regression analysis (see Chapter 7 for a discussion on regression analysis) was performed to determine the interactions of traffic loadings, material properties, layer thickness, and climate. The relationship between axle loads and the thickness index of the pavement system is shown in Fig. 4.7. The thickness index represents a combination of layer thickness and strength coefficients. The term *thickness index* is the same as the term *structural number*, which will be discussed later.

The relationship shown in Fig. 4.7 can be used to determine the thickness of a flexible pavement. For example, assume that a new pavement must withstand 1 million applications of a 24-kip tandem axle load. Based on the curves, it can

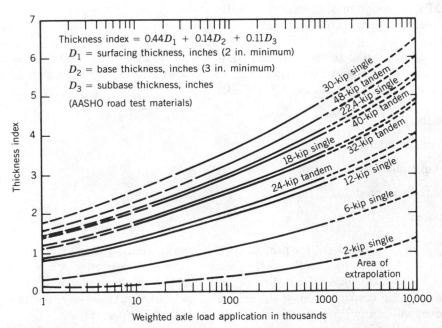

FIGURE 4.7
AASHO road test thickness index versus axle loads. Redrawn from "AASHO Road Test Report 5", Highway Research Board, Special Report 61E, Washington, D.C., 1962. Used by permission.

be seen that a thickness index of approximately 3.0 is needed to sustain that type of loading. There are many combinations of pavement materials and thickness that will provide an index value of 3.0; however, it is the responsibility of the pavement engineer to select a practical and economic combination of the materials to satisfy the design inputs. Since there is an infinite number of combinations of material properties and thicknesses, the graph shown in Fig. 4.7 can quickly lose its utility. Consequently, an equation was developed for flexible pavement design, which replaces the graph in Fig. 4.7.

The basic equation for flexible pavement design given in the 1986 Design Guide permits the engineers to determine a structural number required to carry the design traffic loading. The AASHTO equation is

$$\log_{10} W_{18} = Z_r S_o + 9.36(\log_{10}(SN + 1)) - 0.20 + \frac{\log_{10}[\Delta PSI/(4.2 - 1.5)]}{0.40 + [1094/(SN + 1)^{5.19}]}$$
$$+ 2.32\log_{10} M_R - 8.07 \tag{4.7}$$

where W_{18} = 18-kip equivalent single axle load
Z_r = reliability
S_o = overall standard deviation
SN = structural number
ΔPSI = design present serviceability loss
M_R = resilient modulus of the subgrade soil

The variables that serve as input to Eq. 4.7 are discussed in the following paragraphs. The graphical solution to Eq. 4.7 is shown in Fig. 4.8. The inputs to the nomograph solution (or Eq. 4.7) include:

Reliability.

Overall standard deviation.

18-kip equivalent axle loads.

Soil resilient modulus.

Serviceability loss.

Reliability is defined as "the probability that serviceability will be maintained at adequate levels from a user's point of view, throughout the design life of the facility." This factor insures that the pavement will perform at or above the TSI level during the design period. If the engineer would like to make sure that the pavement will not fail, a high reliability level is selected (90 percent, etc.).

The overall standard deviation factor takes into account the designers ability to estimate the variation in the ability to estimate future 18-kip equivalent axle loads. Typical values for the range of standard deviation for flexible pavement design are 0.4 to 0.5.

$$\log_{10} W_{18} = z_R {}^*S_o + 9.36 {}^*\log_{10}(SN + 1) - 0.20 + \frac{\log_{10}\left[\dfrac{\Delta PSI}{4.2 - 1.5}\right]}{0.40 + \dfrac{1094}{(SN + 1)^{5.19}}} + 2.32 {}^*\log_{10} M_R - 8.07$$

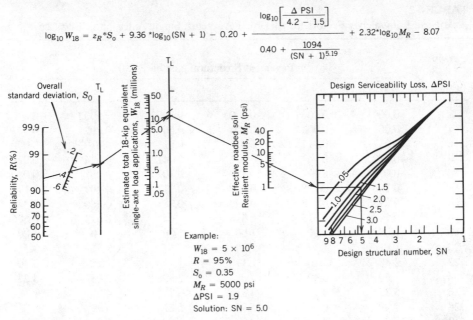

Example:
$W_{18} = 5 \times 10^6$
$R = 95\%$
$S_o = 0.35$
$M_R = 5000$ psi
$\Delta PSI = 1.9$
Solution: SN = 5.0

FIGURE 4.8
Design chart for flexible pavements based on using mean vlaues for each input. Redrawn from "AASHTO Guide for Design of Pavement Structure," Washington, D.C., The American Association of State Highway and Transportation Officials, copyright 1986. Used by permission.

There is a wide range of vehicle loadings on the highway system. Automobile and truck traffic provide a mixed spectrum of axle types and axle loads. To try and design for the mixed traffic loadings, this input variable would require a significant amount of data collection and design evaluation. Instead, the problem of handling mixed traffic loading was solved with the adoption of a standard 18-kip equivalent single-axle load, ESAL. With this method of traffic analysis, the various axle types and loadings can be related to one single-design load.

It was found at the AASHO Road Test that the 18-kip equivalent axle load was also a function of the Terminal Serviceability Index of the pavement structure. The axle load equivalency factors for flexible pavement design, with a TSI of 2.5, are presented in Tables 4.2 and 4.3.

The soil resilient modulus, M_R, is used to reflect the engineering properties of the subgrade. The resilient modulus can be determined by AASHTO Test Method T274. The M_R is important for the design, since it reflects the strength of the soil—which must ultimately support the load. The measurement of the resilient modulus is not performed by all transportation agencies, therefore, a relationship between M_R and the California Bearing Ratio, CBR, has been

TABLE 4.2
Axle Load Equivalency Factors for Flexible Pavements, Single Axles, and TSI = 2.5

Axle Load (kips)	Pavement Structural Number (SN)					
	1	2	3	4	5	6
2	0.0004	0.0004	0.0003	0.0002	0.0002	0.0002
4	0.003	0.004	0.004	0.003	0.002	0.002
6	0.011	0.017	0.017	0.013	0.010	0.009
8	0.032	0.047	0.051	0.041	0.034	0.031
10	0.078	0.102	0.118	0.102	0.088	0.080
12	0.168	0.198	0.229	0.213	0.189	0.176
14	0.328	0.358	0.399	0.388	0.360	0.342
16	0.591	0.613	0.646	0.645	0.623	0.606
18	1.00	1.00	1.00	1.00	1.00	1.00
20	1.61	1.57	1.49	1.47	1.51	1.55
22	2.48	2.38	2.17	2.09	2.18	2.30
24	3.69	3.49	3.09	2.89	3.03	3.27
26	5.33	4.99	4.31	3.91	4.09	4.48
28	7.49	6.98	5.90	5.21	5.39	5.98
30	10.3	9.5	7.9	6.8	7.0	7.8
32	13.9	12.8	10.5	8.8	8.9	10.0
34	18.4	16.9	13.7	11.3	11.2	12.5
36	24.0	22.0	17.7	14.4	13.9	15.5
38	30.9	28.3	22.6	18.1	17.2	19.0
40	39.3	35.9	28.5	22.5	21.1	23.0
42	49.3	45.0	35.6	27.8	25.6	27.7
44	61.3	55.9	44.0	34.0	31.0	33.1
46	75.5	68.8	54.0	41.4	37.2	39.3
48	92.2	83.9	65.7	50.1	44.5	46.5
50	112.0	102.0	79.0	60.0	53.0	55.0

determined, since many agencies use the CBR test to determine the supporting characteristics of the subgrade. The relationship is

$$M_R = 1500 \times \text{CBR} \tag{4.8}$$

The amount of serviceability loss can be determined by the design engineer during the pavement selection process. The engineer must decide the final PSI level for the particular pavement. If the design is for a pavement with heavy traffic load, then the loss may only be 1.2 while a low volume road may have a total loss of 2.7.

TABLE 4.3
Axle Load Equivalency Factors for Flexible Pavements, Tandem Axles and TSI = 2.5

Axle Load (kips)	Pavement Structural Number (SN)					
	1	2	3	4	5	6
2	0.0001	0.0001	0.0001	0.0000	0.0000	0.0000
4	0.0005	0.0005	0.0004	0.0003	0.0003	0.0002
6	0.002	0.002	0.002	0.001	0.001	0.001
8	0.004	0.006	0.005	0.004	0.003	0.003
10	0.008	0.013	0.011	0.009	0.007	0.006
12	0.015	0.024	0.023	0.018	0.014	0.013
14	0.026	0.041	0.042	0.033	0.027	0.024
16	0.044	0.065	0.070	0.057	0.047	0.043
18	0.070	0.097	0.109	0.092	0.077	0.070
20	0.107	0.141	0.162	0.141	0.121	0.110
22	0.160	0.198	0.229	0.207	0.180	0.166
24	0.231	0.273	0.315	0.292	0.260	0.242
26	0.327	0.370	0.420	0.401	0.364	0.342
28	0.451	0.493	0.548	0.534	0.495	0.470
30	0.611	0.648	0.703	0.695	0.658	0.633
32	0.813	0.843	0.889	0.887	0.857	0.834
34	1.06	1.08	1.11	1.11	1.09	1.08
36	1.38	1.38	1.38	1.38	1.38	1.38
38	1.75	1.73	1.69	1.68	1.70	1.73
40	2.21	2.16	2.06	2.03	2.08	2.14
42	2.76	2.67	2.49	2.43	2.51	2.61
44	3.41	3.27	2.99	2.88	3.00	3.16
46	4.18	3.98	3.58	3.40	3.55	3.79
48	5.08	4.80	4.25	3.98	4.17	4.49
50	6.12	5.76	5.03	4.64	4.86	5.28
52	7.33	6.87	5.93	5.38	5.63	6.17
54	8.72	8.14	6.95	6.22	6.47	7.15
56	10.3	9.6	8.1	7.2	7.4	8.2
58	12.1	11.3	9.4	8.2	8.4	9.4
60	14.2	13.1	10.9	9.4	9.6	10.7
62	16.5	15.3	12.6	10.7	10.8	12.1
64	19.1	17.6	14.5	12.2	12.2	13.7
66	22.1	20.3	16.6	13.8	13.7	15.4
68	25.3	23.3	18.9	15.6	15.4	17.2
70	29.0	26.6	21.5	17.6	17.2	19.2
72	33.0	30.3	24.4	19.8	19.2	21.3
74	37.5	34.4	27.6	22.2	21.3	23.6
76	42.5	38.9	31.1	24.8	23.7	26.1
78	48.0	43.9	35.0	27.8	26.2	28.8
80	54.0	49.4	39.2	30.9	29.0	31.7
82	60.6	55.4	43.9	34.4	32.0	34.8
84	67.8	61.9	49.0	38.2	35.3	38.1
86	75.7	69.1	54.5	42.3	38.8	41.7
88	84.3	76.9	60.6	46.8	42.6	45.6
90	93.7	85.4	67.1	51.7	46.8	49.7

TABLE 4.4
Structural Layer Coefficients

Pavement Component	Coefficient
Surface Course	
Hot mix asphalt concrete	0.44
Sand mix asphalt concrete	0.35
Base Course	
Crushed stone	0.14
Dense graded crushed stone	0.18
Soil cement	0.20
Emulsion/aggregate-bituminous	0.30
Portland cement/aggregate	0.40
Lime-pozzolan/aggregate	0.40
Hot mix asphaltic concrete	0.40
Subbase	
Crushed stone	0.11

Structural Number

After the input variables have been selected and applied to Fig. 4.8, a design structural number, SN, is selected. The SN represents the overall pavement system structural requirements needed to sustain the design traffic loadings for the design period. The structural number is similar to the Thickness Index as shown in Fig. 4.7. As stated earlier there is a number of pavement material combinations and thickness that will provide satisfactory service. The following equation can be used to relate individual material types and thickness to the structural number. Typical layer coefficients are shown in Table 4.4.

$$SN = a_1 D_1 + a_2 D_2 M_2 + a_3 D_3 M_3 \qquad (4.9)$$

where a_1, a_2, a_3 = layer coefficient of the wearing, base and subbase layer, respectively

D_1, D_2, D_3 = thickness of wearing, base and subbase, respectively

M_2, M_3 = drainage coefficient for the base and subbase, respectively

A drainage coefficient, M_i, is used to modify the thickness of the lower pavement layers to take into account drainage characteristics. A value of 1.0 for M_i represents good drainage characteristics. A soil such as clay will not drain very well; consequently, it will have a lower coefficient. Since there are many combinations of layer coefficient and thickness that solve Eq. 4.9, there are some guidelines that can be used to narrow the number of solutions. Experience has shown that the wearing layer can be 2 to 4 in. thick while the subbase can be 4 to 8 in. thick. Knowing which of the materials is the most costly per inch will also assist with the solution of an initial layer thickness.

Example 4.2

Using the AASHTO Flexible Pavement Design Procedure, we select a pavement cross section that will provide 10 years service. The initial PSI is 4.2. We assume that the final PSI is 2.5. The subgrade CBR is 12. Reliability is 95 percent with a standard deviation of 0.4. The daily truck traffic consists of 400 passes of trucks with two single axles and 350 passes of semitrailer truck with tandem axles. The axle weights are

$$\text{single-unit truck} = 8000\text{-lb steering, single axle}$$
$$22,000\text{-lb drive, single axle}$$

$$\text{semiunit truck} = 10,000\text{-lb steering, single axle}$$
$$= 16,000\text{-lb drive, tandem axle}$$
$$= 34,000\text{-lb trailer, tandem axle}$$

M_1 and M_2 are equal to 1.0 for the materials in the pavement.

Solution

Determine inputs for Fig. 4.8.
   Roadbed soil resilient modulus:

$$M_R = 1500\ \text{CBR}$$
$$M_R = 1500(12)$$
$$M_R = 18,000\ \text{psi}$$

18-kip-ESAL calculation

An assumption for the SN is needed to start the problem. It is used to select a column for Tables 4.2 and 4.3. An initial SN = 4 is usually appropriate.
   For the single-unit truck (Table 4.2):

$$8000\text{-lb axle equivalent} = 0.041$$

$$22,000\text{-lb axle equivalent} = \underline{2.090}$$

$$\text{ESAL for the truck} = 2.131$$

For the semiunit truck (Tables 4.2 and 4.3):

$$10,000\text{-lb axle equivalent} = 0.102$$
$$16,000\text{-lb axle equivalent} = 0.057$$
$$34,000\text{-lb axle equivalent} = \underline{1.110}$$
$$\text{ESAL for the truck} = 1.269$$

The truck traffic for the design period is

$$400 \text{ passes} \times 2.131 = 852.4$$
$$350 \text{ passes} \times 1.269 = 444.2$$
$$\text{total } 18\text{-}k \text{ ESAL/day} = 1297$$

Traffic for the design period (10 years):

$$1297 \times 365 \times 10 = 4{,}732{,}407 \text{ 18-kip-ESAL}$$

With an initial PSI of 4.2 and a TSI of 2.5, the Δ equals

$$4.2 - 2.5 = 1.7$$

Using Fig. 4.8, we find that the SN is 3.4. Using Equation 4.9, we obtain

$$SN = a_1 D_1 + a_2 D_2 M_2 + a_3 D_3 M_3$$

since M_2 and $M_3 = 1.0$.

Select a wearing course of hot mix asphalt concrete:

$$a_1 = 0.44 \qquad D_1 = 3.0 \text{ in.}$$

Select a base course of crushed stone:

$$a_2 = 0.14 \qquad D_2 = ?$$

Select a subbase course:

$$a_3 = 0.11$$
$$D_3 = 6.0 \text{ in.}$$

Then

$$3.4 = 0.44(3.0) + 0.14(D_2)(1.0) + 0.11(6.0)(1.0)$$
$$D_2 = 10.1 \text{ in., say } 10.0 \text{ in.}$$

The final cross section can be

3.0 in. of asphalt concrete wearing

10.0 in. of crushed aggregate base

6.0 in. of crushed aggregate subbase

4.5 PAVEMENT SYSTEM DESIGN PRINCIPLES FOR RIGID PAVEMENTS

A rigid pavement distributes traffic loads to the subgrade by the beam action of the Portland cement concrete (PCC) slab. The PCC slab material has a modulus of elasticity of approximately 4.5 million psi, which is much greater than the modulus of elasticity for the flexible pavement materials. Because of this beam action, there is a much larger area of load distribution than with a flexible pavement; thus, the need for a different type of analysis.

4.5.1 Calculation of Rigid Pavement Stresses and Deflections

H. M. Westergaard presented a theoretical analysis for rigid pavements in 1925 [1926]. He assumed that the PCC slab acts as a homogenous, isotropic, and elastic solid. He also assumed that the subgrade reacts much like a liquid such that the deflections of the slab at any point are a function of the load and the modulus of the subgrade reaction, k, as shown by

$$p = k\Delta \tag{4.10}$$

where k is the modulus of subgrade reaction in lb/in.^{3}, p is the reactive pressure in psi, and Δ is the slab deflection in inches. The modulus k is assumed to be constant at each point under the slab and independent of deflection. The Westergaard equations were developed for three loading cases: an interior load, an edge load, and a corner load, as shown in Fig. 4.9. The equations have been referenced throughout the pavement design literature to the point where some equations have been miscopied and misused. Even Dr. Westergaard had to correct his equations in later publications. Recently, Ioannides, Thompson, and Barenberg reconsidered the Westergaard solutions and compared them with finite element analysis [1985]. Their work has refined and corrected the original solutions. These closed-form solutions provide the foundation for rigid pavement analysis.

The Westergaard solutions for stresses and deflections as presented by Ioannides, Thompson and Barenberg are as follows.

For the interior loading:

$$\sigma_i = \frac{3P(1+\mu)}{2\pi h^2}\left(\ln\left(\frac{2l}{b}\right) + 0.5 - \gamma\right) + \frac{3P(1+\mu)}{64h^2}\left(\frac{b}{l}\right)^2 \tag{4.11}$$

$$\Delta_i = \frac{P}{8kl^2}\left(1 + \left(\frac{1}{2\pi}\right)\left(\ln\left(\frac{a}{2l}\right) + \gamma - \frac{5}{4}\right)\left(\frac{a}{l}\right)^2\right) \tag{4.12}$$

where σ_i = bending stress, psi
P = total load, lb
E = modulus of elasticity, psi
μ = Poisson's ratio
h = slab thickness, in

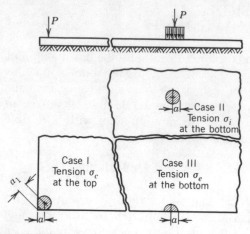

FIGURE 4.9
Westergaard loading cases. Redrawn from "Computation of Stresses in Concrete Roads", by H. M. Westergaard, Highway Research Board Proceedings of the Fifth Annual Meeting, Washington, D.C., 1926. Used by permission.

k = modulus of subgrade reaction, pci
a = radius of circular load

$$l^4 = \frac{Eh^3}{12(1 - \mu^2)k} \quad \text{which is the radius of relative stiffness} \tag{4.13}$$

$$b = (1.6a^2 + h^2)^{0.5} - 0.675h \quad \text{if } a < 1.724h \tag{4.14}$$

or

$$b = a \quad \text{if } a > 1.724h$$
$$\gamma = \text{Euler's constant} = 0.577215$$

For the edge loading:

$$\sigma_e = 0.529(1 + 0.54\mu)\left(\frac{P}{h^2}\right)\left[\log_{10}\left(\frac{Eh^3}{ka^4}\right) - 0.71\right] \tag{4.15}$$

$$\Delta_e = 0.408(1 + 0.4\mu)\left(\frac{P}{kl^2}\right) \tag{4.16}$$

For the corner loading:

$$\sigma_c = \frac{3P}{h^2}\left[1 - \left(\frac{a_1}{l}\right)^{0.6}\right] \tag{4.17}$$

$$\Delta_c = \frac{P}{kl^2}\left[1.1 - 0.88\left(\frac{a_l}{l}\right)\right] \tag{4.18}$$

where a_1 = distance to point of action of resulted load on common angle bisection.

Example 4.3

A 30,000-lb single axle load is placed on a PCC slab that is 10.0 in. thick. The concrete has a modulus of elasticity of 4.5 million psi with a Poisson's ratio of 0.18. The modulus of subgrade reaction is 200 pci. Tire pressure is 100 psi and a_1 is 12.0 in. Calculate the stress and deflection if the load is placed on the corner of the slab.

Solution

The Westergaard equation for corner stress (Eq. 4.17) is

$$\sigma_c = \frac{3P}{h^2}\left[1 - \left(\frac{a_1}{l}\right)^{0.6}\right]$$

Calculate l, the radius of relative stiffness.

$$l^4 = \frac{Eh^3}{12(1 - \mu^2)k}$$

$$l^4 = \frac{4,500,000(10)^3}{12(1 - (0.18)^2)200}$$

$$l = 37.06$$

Substituting terms, we obtain

$$\sigma_c = \frac{3(15,000)}{(10)^2}\left[1 - \left(\frac{12}{37.06}\right)^{0.6}\right]$$

$$= 220.5 \text{ psi}$$

The Westergaard equation for corner deflection is

$$\Delta_c = \frac{P}{kl^2}\left[1.1 - 0.88\left(\frac{a_1}{l}\right)\right]$$

Substituting gives

$$\Delta_c = \frac{15,000}{200(37.06)^2}\left[1.1 - 0.88\left(\frac{12}{37.06}\right)\right]$$

$$= 0.045 \text{ in.}$$

4.6 RIGID PAVEMENT DESIGN PROCEDURE

The design of a rigid pavement is similar in concept to the design of a flexible pavement system. The major difference between the flexible and rigid pavements is the failure mode. A flexible pavement usually fails due to rutting and fatigue cracking (alligator cracking). A rigid pavement, however, usually fails due to joint and crack deterioration. Transverse and longitudinal cracks as well as corner breaks also lead to a failed rigid pavement system.

4.6.1 The AASHTO Rigid Pavement Design Procedure

The AASHTO Design Guide can be used for the design of a rigid pavement system. The design procedure is based on the field results of the AASHO Road Test. The procedure is applicable to jointed plain (JPCP), reinforced (JRCP), and continuous reinforced pavements (CRCP). The JPCP does not have any steel reinforcement in the slab, while the JRCP has welded wire fabric. The CRCP has steel bars for reinforcement. Since faulting, which is a distress due to different slab elevations, was not a failure mechanism at the AASHO Road Test, the design of nondowelled joints must be checked with another design procedure.

The design procedure for rigid pavements is based on a selected reduction in serviceability. The regression equation that can be used to determine the rigid pavement thickness is

$$\log_{10} W_{18} = Z_R * S_o + 7.35 * \log_{10}(D+1) - 0.06 + \frac{\log_{10}[\Delta PSI/(4.5-1.5)]}{1 + [1.624*10^7/(D+1)^{8.46}]}$$

$$+ (4.22 - 0.32 TSI) * \log_{10} \frac{S_c' * C_d[D^{0.75} - 1.132]}{215.63 * J[D^{0.75} - [18.42/(E_c/k^{0.25})]]} \quad (4.19)$$

where W_{18} = 18-kip equivalent single axle load
   Z_R = reliability
   S_o = overall standard deviation
   S_c' = concrete modulus of rupture
   J = load transfer coefficient
   E_c = concrete modulus of elasticity
   k = modulus of subgrade reaction
   D = slab thickness

The design equation explains the relationship between 18-kip traffic, thickness, and loss of serviceability. It can be solved with a computer or with a nomograph as shown in Figs. 4.10 and 4.11. The inputs to the equation or nomograph include:

Estimated future traffic.

Reliability.

Overall standard deviation.

Nomograph solves:

$$\log_{10} W_{18} = Z_R {}^* S_o + 7.35 {}^* \log_{10}(D + 1) - 0.06 + \frac{\log_{10}\left[\dfrac{\Delta \text{PSI}}{4.5 - 1.5}\right]}{1 + \dfrac{1.624{}^* 10^7}{(D + 1)^{8.46}}} + (4.22 - 0.32_{TSI}){}^* \log_{10}\left[\frac{s'_c {}^* c_d \left[D^{0.75} - 1.132\right]}{215.63{}^* J \left[D^{0.75} - \dfrac{18.42}{(E_c/k)^{0.25}}\right]}\right]$$

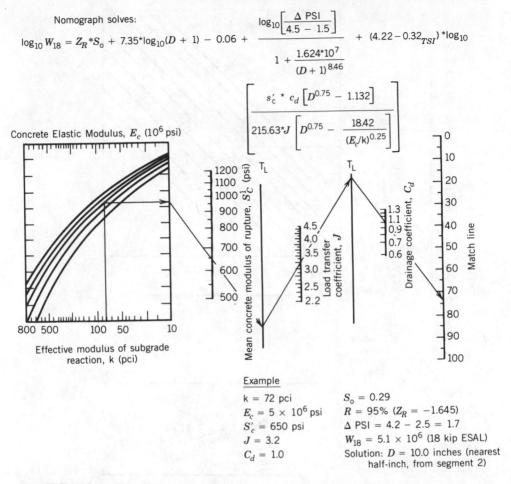

FIGURE 4.10
Segment 1 of the design chart for rigid pavement based on using mean values for each input variable. Redrawn from "AASHTO Guide for Design of Pavement Structures," The American Association of State Highway and Transportation Officials, Washington, D.C., copyright 1986. Used by Permission.

Serviceability loss.

Concrete elastic modulus.

Concrete modulus of rupture.

Load transfer coefficient.

Several of the input parameters have been discussed under the flexible pavement design procedure section. Variables that are different for the rigid pavement design procedure are explained in the following paragraphs.

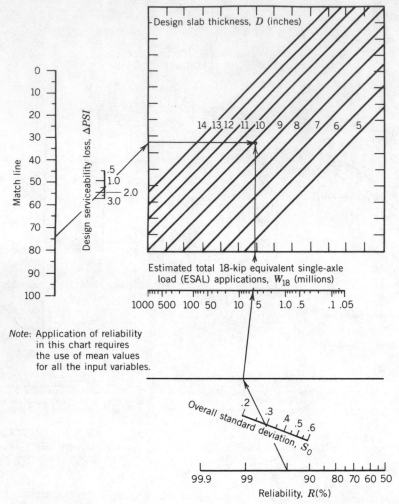

FIGURE 4.11
Segment 2 of the design chart for rigid pavements based on using mean values for each input variable. *Note*: Application of reliability in this chart requires the use of mean values for all the input variables. Redrawn from "AASHTO Guide for Design of Pavement Structure," The American Association of State Highway and Transportation Officials, Washington, D.C., copyright 1986. Used by permission.

The concrete modulus of elasticity is derived from the stress strain curve as taken in the elastic region. As discussed previously, the modulus of elasticity is also known as Young's modulus. Portland cement concrete can have an elastic modulus between 3 and 7 million psi.

The effective modulus of subgrade reaction, k, depends upon several different factors including moisture content and density. The k value can be estimated by the following formula.

$$k = \frac{M_R}{19.4} \tag{4.20}$$

Typical values for k can have a range of 10 to 800 pci.

TABLE 4.5
Axle Load Equivalency Factors for Rigid Pavements, Single Axles and TSI = 2.5

Axle Load (kips)	Slab Thickness, D (inches)								
	6	7	8	9	10	11	12	13	14
2	0.0002	0.0002	0.0002	0.0002	0.0002	0.0002	0.0002	0.0002	0.0002
4	0.003	0.002	0.002	0.002	0.002	0.002	0.002	0.002	0.002
6	0.012	0.011	0.010	0.010	0.010	0.010	0.010	0.010	0.010
8	0.039	0.035	0.033	0.032	0.032	0.032	0.032	0.032	0.032
10	0.097	0.089	0.084	0.082	0.081	0.080	0.080	0.080	0.080
12	0.203	0.189	0.181	0.176	0.175	0.174	0.174	0.174	0.173
14	0.376	0.360	0.347	0.341	0.338	0.337	0.336	0.336	0.336
16	0.634	0.623	0.610	0.604	0.601	0.599	0.599	0.599	0.598
18	1.00	1.00	1.00	1.00	1.00	1.00	1.00	1.00	1.00
20	1.51	1.52	1.55	1.57	1.58	1.58	1.59	1.59	1.59
22	2.21	2.20	2.28	2.34	2.38	2.40	2.41	2.41	2.41
24	3.16	3.10	3.22	3.36	3.45	3.50	3.53	3.54	3.55
26	4.41	4.26	4.42	4.67	4.85	4.95	5.01	5.04	5.05
28	6.05	5.76	5.92	6.29	6.61	6.81	6.92	6.98	7.01
30	8.16	7.67	7.79	8.28	8.79	9.14	9.35	9.46	9.52
32	10.8	10.1	10.1	10.7	11.4	12.0	12.3	12.6	12.7
34	14.1	13.0	12.9	13.6	14.6	15.4	16.0	16.4	16.5
36	18.2	16.7	16.4	17.1	18.3	19.5	20.4	21.0	21.3
38	23.1	21.1	20.6	21.3	22.7	24.3	25.6	26.4	27.0
40	29.1	26.5	25.7	26.3	27.9	29.9	31.6	32.9	33.7
42	36.2	32.9	31.7	32.2	34.0	36.3	38.7	40.4	41.6
44	44.6	40.4	38.8	39.2	41.0	43.8	46.7	49.1	50.8
46	54.5	49.3	47.1	47.3	49.2	52.3	55.9	59.0	61.4
48	66.1	59.7	56.9	56.8	58.7	62.1	66.3	70.3	73.4
50	79.4	71.7	68.2	67.8	69.6	73.3	78.1	83.0	87.1

TABLE 4.6
Axle Load Equivalency Factors for Rigid Pavements, Tandem Axles and TSI = 2.5

Axle Load (kips)	Slab Thickness, D (inches)								
	6	7	8	9	10	11	12	13	14
2	0.0001	0.0001	0.0001	0.0001	0.0001	0.0001	0.0001	0.0001	0.0001
4	0.0006	0.0006	0.0005	0.0005	0.0005	0.0005	0.0005	0.0005	0.0005
6	0.002	0.002	0.002	0.002	0.002	0.002	0.002	0.002	0.002
8	0.007	0.006	0.006	0.005	0.005	0.005	0.005	0.005	0.005
10	0.015	0.014	0.013	0.013	0.012	0.012	0.012	0.012	0.012
12	0.031	0.028	0.026	0.026	0.025	0.025	0.025	0.025	0.025
14	0.057	0.052	0.049	0.048	0.047	0.047	0.047	0.047	0.047
16	0.097	0.089	0.084	0.082	0.081	0.081	0.080	0.080	0.080
18	0.155	0.143	0.136	0.133	0.132	0.131	0.131	0.131	0.131
20	0.234	0.220	0.211	0.206	0.204	0.203	0.203	0.203	0.203
22	0.340	0.325	0.313	0.308	0.305	0.304	0.303	0.303	0.303
24	0.475	0.462	0.450	0.444	0.441	0.440	0.439	0.439	0.439
26	0.644	0.637	0.627	0.622	0.620	0.619	0.618	0.618	0.618
28	0.855	0.854	0.852	0.850	0.850	0.850	0.849	0.849	0.849
30	1.11	1.12	1.13	1.14	1.14	1.14	1.14	1.14	1.14
32	1.43	1.44	1.47	1.49	1.50	1.51	1.51	1.51	1.51
34	1.82	1.82	1.87	1.92	1.95	1.96	1.97	1.97	1.97
36	2.29	2.27	2.35	2.43	2.48	2.51	2.52	2.52	2.53
38	2.85	2.80	2.91	3.03	3.12	3.16	3.18	3.20	3.20
40	3.52	3.42	3.55	3.74	3.87	3.94	3.98	4.00	4.01
42	4.32	4.16	4.30	4.55	4.74	4.86	4.91	4.95	4.96
44	5.26	5.01	5.16	5.48	5.75	5.92	6.01	6.06	6.09
46	6.36	6.01	6.14	6.53	6.90	7.14	7.28	7.36	7.40
48	7.64	7.16	7.27	7.73	8.21	8.55	8.75	8.86	8.92
50	9.11	8.50	8.55	9.07	9.68	10.14	10.42	10.58	10.66
52	10.8	10.0	10.0	10.6	11.3	11.9	12.3	12.5	12.7
54	12.8	11.8	11.7	12.3	13.2	13.9	14.5	14.8	14.9
56	15.0	13.8	13.6	14.2	15.2	16.2	16.8	17.3	17.5
58	17.5	16.0	15.7	16.3	17.5	18.6	19.5	20.1	20.4
60	20.3	18.5	18.1	18.7	20.0	21.4	22.5	23.2	23.6
62	23.5	21.4	20.8	21.4	22.8	24.4	25.7	26.7	27.3
64	27.0	24.6	23.8	24.4	25.8	27.7	29.3	30.5	31.3
66	31.0	28.1	27.1	27.6	29.2	31.3	33.2	34.7	35.7
68	35.4	32.1	30.9	31.3	32.9	35.2	37.5	39.3	40.5
70	40.3	36.5	35.0	35.3	37.0	39.5	42.1	44.3	45.9
72	45.7	41.4	39.6	39.8	41.5	44.2	47.2	49.8	51.7
74	51.7	46.7	44.6	44.7	46.4	49.3	52.7	55.7	58.0
76	58.3	52.6	50.2	50.1	51.8	54.9	58.6	62.1	64.8
78	65.5	59.1	56.3	56.1	57.7	60.9	65.0	69.0	72.3
80	73.4	66.2	62.9	62.5	64.2	67.5	71.9	76.4	80.2
82	82.0	73.9	70.2	69.6	71.2	74.7	79.4	84.4	88.8
84	91.4	82.4	78.1	77.3	78.9	82.4	87.4	93.0	98.1
86	102.0	92.0	87.0	86.0	87.0	91.0	96.0	102.0	108.0
88	113.0	102.0	96.0	95.0	96.0	100.0	105.0	112.0	119.0
90	125.0	112.0	106.0	105.0	106.0	110.0	115.0	123.0	130.0

The concrete modulus of rupture, S'_c, is a measure of the tensile strength of the concrete. It is determined by loading a beam specimen, at the third points, to failure. Typical values are 500 to 1200 psi.

The load transfer coefficient, J, is a factor that is used to account for the ability of the pavement to transfer load from one slab to another across a joint or crack. A typical value for J is 3.2.

Inputs for reliability, overall standard deviation, and 18-kip ESALs are similar to the factors for flexible pavement design procedure. The only differences are that the 18-kip ESAL factors for rigid pavements are different from the flexible pavement factors. Also, the drainage coefficients are slightly different from flexible pavements, but a value for good drainage is 1.0. The 18-kip ESAL factors for rigid pavements are shown in Tables 4.5 and 4.6.

Example 4.4

Using the AASHTO Rigid Pavement Design Procedure, we select a slab thickness that will provide for 20 years service. Assume that the final PSI is 2.5. The modulus of subgrade reaction is 400. Daily truck traffic consists of 200 passes of a truck with two single axles and 375 passes of a semitrailer with a tandem axle. The axle weights are

Single-unit truck: 10,000-lb steering, single axle

24,000-lb drive, single axle.

Semiunit truck: 8,000-lb steering, single axle

20,000-lb drive, tandem axle

32,000-lb trailer, tandem axle.

Reliability = 95 percent.

Standard deviation = 0.35.

Modulus of elasticity (concrete) = 4.5 million psi.

Modulus of rupture, S'_c = 700 psi.

J = 3.2 and drainage coefficient = 1.0.

Solution

The first step is to determine the inputs for Fig. 4.10.
18-kip ESAL calculation

Assume that $D = 10.0$ in. (This is needed to begin the problem and select values from Tables 4.4 and 4.5.)

For single-unit truck (Table 4.4):

$$10,000\text{-lb axle equivalent} = 0.081$$
$$24,000\text{-lb axle equivalent} = \underline{3.450}$$
$$\text{ESAL for truck} = 3.531$$

For semiunit truck (Tables 4.4 and 4.5):

$$8,000\text{-lb axle equivalent} = 0.032$$
$$20,000\text{-lb axle equivalent} = 0.204$$
$$32,000\text{-lb axle equivalent} = \underline{1.500}$$
$$\text{ESAL for truck} = 1.736$$

The truck traffic for the design period is

$$200 \text{ passes} \times 3.531 = 706.2$$
$$375 \text{ passes} \times 1.736 = 651.0$$
$$\text{Total 18-kip ESAL/day} = 1357$$

The traffic for the design period (20 yr) is

$$1357 \times 365 \times 20 = 9,907,560 \quad \text{18-kip ESAL}$$

If we assume that an initial PSI of 4.2 and a TSI of 2.5, the ΔPSI equals

$$4.2 - 2.5 = 1.7$$

Using Fig. 4.10, we find that $D = 9.75$ in. This will probably be rounded to 10.0 in. for construction purposes.

NOMENCLATURE
FOR
CHAPTER 4

a equivalent tire radius

a_i structural layer coefficient

CBR   California bearing ratio

D slab thickness, AASHTO design equation

D_i	layer thickness, flexible pavement
E	Young's modulus
h	slab thickness, Westergaard solutions
l	radius of relative stiffness
k	modulus of subgrade reaction
m_i	drainage coefficient
pci	pounds per cubic inch
p	tire pressure
P	wheel load
psi	pounds per square inch
PSI	present serviceability index
S_c'	concrete modulus of rupture
S_o	standard deviation for AASHTO design equations
SN	structural number
TSI	terminal serviceability index
W_{18}	18-kip equivalent single-axle load
z	depth in pavement
Z_i	reliability for AASHTO design equations
Δ_c	corner deflection
Δ_e	edge deflection
Δ_i	deflection
Δ_{PSI}	lose of present serviceability index
σ_c	bending stress, corner
σ_e	bending stress, edge
σ_i	bending stress, interior
σ_r	horizontal radial stress
σ_z	vertical stress
μ	Poisson ratio
γ	Euler constant

REFERENCES

1. E. J. Yoder, and M. W. Witczak, *Principles of Pavement Design*, 2nd ed., John Wiley & Sons, 1975.

2. R. G. Ahlvin, and H. H. Ulery, "Tabulated Values for Determining the Complete Pattern of Stresses, Strains, and Deflections Beneath a Uniform Circular Load on a Homogeneous Half Space," Highway Research Board Bulletin 342, 1962.

3. W. Carey, and P. Irick, "The Pavement Serviceability-Performance Concept," Highway Research Board Special Report 61E, AASHO Road Test, 1962.

4. H. M. Westergaard, "Computation of Stresses in Concrete Roads," Proceedings of the Fifth Annual Meeting of the Highway Research Board, Washington, D.C., 1926.

5. A. M. Ioannides, M. R. Thompson, and E. J. Barenberg, "Westergaard Solutions Reconsidered," Transportation Research Record 1043, Washington, D.C., 1985.

6. "AASHTO Guide for Design of Pavement Structures," The American Association of State Highway and Transportation Officials, Washington, D.C., 1986.

7. "AASHTO Road Test Report 5," Highway Research Board, Special Report G, Washington, D.C., 1962.

8. H. M. Westergaard, "Computation of Stresses in Concrete Roads," Proceedings of the Fifth Annual Meeting of the Highway Research Board, Washington, D.C., 1927.

PROBLEMS

4.1. One tire on a vehicle has a tire pressure of 100 psi with a contact area of 78.5 sq in. The pavement material has an $E = 50,000$ psi and a Poisson ratio of 0.45. Determine the radial horizontal stress and the deflection at

(a) A point under the center of the load at the pavement surface.

(b) A point 20 in. deep at a radial distance of 10 in.

4.2. A 26,000-lb single-axis truck has a tire pressure of 80 psi. The load is placed on a 9.0-in. PCC slab with a modulus of 4,000,000 psi and a Poisson ratio of 0.15. The subgrade k is 300 pci. The load is 12.0 in. from the corner. Calculate the edge and corner stresses.

4.3. An 800-lb wheel load is exerted on a pavement by an automobile. The auto's tire pressure is 30 psi, and the surface's modulus of elasticity is 15,000 psi.

(a) What is the pavement surface deflection?

(b) Assuming plate loading, what is the vertical stress imposed by the wheel load at a depth of 10 in.?

4.4. A 10,000-lb wheel load is applied to a pavement and generates a vertical deflection of 0.0061 inches at a depth of 3 in.

(a) If the equivalent radius of tire contact is 4 in., what is the pavement's modulus of elasticity?

(b) If the pavement is 5 in. thick, what is the vertical stress at the top of the subgrade?

4.5. A wheel load produces a surface deflection of 0.0087 in. on a pavement that has a modulus of elasticity of 15,000 psi.

(a) If the tire pressure is 30 psi, how heavy is the wheel load?

(b) If another vehicle exerts a wheel load of 2000 lb and the radius of the tire contact area is 3.57 in., what is the stress 5 in. below the pavement surface?

4.6. If a truck has four tires with a pressure of 80 psi each, and each axle has a load of 22,000 lb, what will the stress be in the sandy soil pavement, below each tire, at (a) the soil pavement surface and (b) 10 in. below the surface?

4.7. A truck (truck A) has two single axles. The steering axle weighs 10,000 lb and the drive axle weighs 23,000 lb. The tires are inflated to 100 psi. A second truck (truck B) has a steering axle that weighs 8000 lb and a tandem axle that weighs 43,000 lb. The soil near Cleveland has a CBR = 10. A pavement in the area had a design structural number of 4.75. The Poisson ratio for concrete is 0.15. The structure value for subbase is 0.11. Which truck will do more damage to the pavement system?

4.8. A highway has the following pavement cross section: 3.0 in. of hot mix, 7.0 in. of crushed stone DG, 10 in. of subbase on a soil with a CBR of 10. How many 25-kip single-axle passes can be carried by the flexible pavement before the serviceability reaches 2.5? Assume that reliability is 90 percent with a standard deviation of 0.40. The resilient modulus is 5000 psi with a ΔPSI of 1.5.

4.9. You have been asked to design the pavement for the access road to a major truck terminal. The daily truck traffic consists of the following:

50 passes of a 39,000-lb tandem axle.
570 passes of a 25,000-lb tandem axle.
80 passes of a 22,500-lb single axle.

There are also 8500 automobiles expected on the roadway every day. The soil has a CBR of 13. The following materials are available: hot mix asphalt, crushed stone, aggregate cement, bituminous concrete, subbase, portland cement concrete (module of rupture = 550 psi). Determine three different pavement cross sections that will structurally support the traffic. Assume that reliability is 95 percent and the standard deviation is 0.4. All soils have a drainage coefficient of 1.0. The ΔPSI is 1.7. J is 3.2. The E for PCC is 5,000,000 psi.

4.10. You have been asked to design a flexible pavement in a city to replace one that disappeared during a recent earthquake. The following traffic is expected:

Daily Count	Axle Load
5000 (single axle)	10,000 lb
1000 (tandem axle)	30,000 lb
400 (single axle)	24,000 lb
100 (tandem axle)	50,000 lb

(a) If the soil CBR is 15, what is the required structural number? (Assume values for reliability, standard deviation, etc.).

(b) Design a pavement that achieves the necessary SN. Show all calculations and materials used.

4.11. Given the same conditions as in Problem 4.10, suppose that you were asked to design a rigid pavement using concrete with a modulus of rupture of 600 psi and $J = 3.2$.

(a) What depth of slab is required? (Assume values for reliability, etc.)

4.12. A pavement is being constructed in Pittsburgh, Pennsylvania, to support the following traffic:

Daily Count	Axle Load
300 (single axle)	10,000 lb
120 (single axle)	18,000 lb
100 (single axle)	23,000 lb
100 (tandem axle)	32,000 lb
30 (single axle)	32,000 lb

A flexible pavement is designed to have 4 in. of hot mix surface course, 6 in. of cement concrete base, and 7 in. of subbase. Reliability is 90 percent with a standard deviation of 0.45. Drainage coefficients are 1.0. The terminal PSI is 2.5.

(a) What is the minimum acceptable soil CBR value?

(b) The state has relaxed weight limits and the impact has been to reduce the number of 18,000-lb single-axle loads from 120 to 20 and increase the number of 32,000-lb single-axle loads from 30 to 90 (all other traffic is unaffected). Under these revised daily counts, what is the minimum acceptable soil CBR value?

4.13. You have been asked by the Soviet Union to design a highway that is to be used in Moscow. The following traffic is expected:

Daily Count	Axle Load
1300 (single axle)	8,000 lb
900 (tandem axle)	15,000 lb
20 (single axle)	40,000 lb
200 (tandem axle)	40,000 lb

A Russian engineer originally designed the road (flexible pavement) with 4 in. of hot mix asphalt surface courses, 4 in. of bituminous concrete base, and 8 in. of subbase. If the reliability was 70 percent and the S_o was 0.5 and ΔPSI was 2.0, what was the CBR value of the soil subgrade?

4.14. Another road (rigid pavement) is being designed with the same loading conditions as Problem 4.12. If the subgrade k is 300 pci and the slab thickness is determined to be 8.5 in., what was the design modulus of rupture?

4.15. A rigid pavement was designed in Pennsylvania. The final design consisted of an 8.5-in. slab on a soil with a k of 400 pci. The modulus of rupture was assumed to be 550 psi. If PennDOT decides to replace the PCC pavement with a flexible pavement, what structural number is required? The TSI is 2.5 with reliability of 90 percent and S_o of 0.45. Also, J is 3.2 and the drainage coefficient is 1.0. Soil modulus = 7000 psi.

4.16. Your engineering firm has been awarded the contract to do the pavement design for a bypass. Provide a thickness of each layer in the cross section for two alternative designs: (a) a rigid pavement and (b) the *minimum* thickness flexible pavement (given available materials). The soil CBR = 12; K = 400 pci.

Daily traffic volumes:
  5000 automobiles.
  300 tractor-trailers (each with one 18K single axle, one 26K tandem axle, and one 36K tandem axle).
Available materials:
  Hot mix asphalt surface course (PennDOT specifies 3 in.).
  Crushed stone base course.
  Aggregate bituminous base course.
  Bituminous concrete base course.
  Standard subbase.
  Portland cement concrete (modulus of rupture = 500 psi).
  Reliability is 95 percent with S_o equal to 0.45.
  TSI is 2.5.

Sketch the cross section of your two designs.

Chapter Five

Elements of Traffic Analysis

5.1 INTRODUCTION

It is important to remember that the primary function of a highway is to provide a transportation service. In an engineering context this service is measured in terms of the ability of a highway to accommodate vehicular traffic safely and efficiently. Thus the basis for determining the functional effectiveness of any highway lies in the vehicular analysis of the traffic.

In undertaking such an analysis, the various dimensions of traffic, such as quantity, type, speed, and distribution over time, must be addressed, since they will influence highway design (selection of number of lanes, pavement types, and geometric design) and highway operations (selection of traffic control devices including signs, markings, and signals), both of which impact functional effectiveness. Therefore, the provision of theoretically consistent quantitative techniques by which relevant dimensions of vehicular traffic can be modeled forms the basis of traffic analysis. This chapter focuses on the presentation of the quantitative elements that are fundamental to contemporary traffic analysis. A thorough understanding of these elements will provide a valuable foundation for the comprehension and critical assessment of the various traffic analysis techniques and procedures that have become the standard of engineering practice.

5.2 TRAFFIC FLOW, SPEED, AND DENSITY

Traffic flow, speed, and density are variables that form the underpinnings of traffic analysis. To begin to study these variables, some basic definitions must first be presented. Traffic flow, q, is defined as the number of vehicles, n, passing some designated highway point in a time interval of duration t, or

$$q = \frac{n}{t} \tag{5.1}$$

where q is generally expressed in vehicles per unit time. Aside from knowing the total number of vehicles arriving in some time interval, the amount of time between the arrival of successive vehicles (or the temporal distribution of traffic flow) is also of interest. The time between the passage of the front bumpers of successive vehicles, past some designated highway point, is known as the time headway. The relationships between time headways, h_i and Eq. 5.1 are

$$t = \sum_{i=1}^{n} h_i \tag{5.2}$$

and

$$q = \frac{n}{\displaystyle\sum_{i=1}^{n} h_i}$$

or

$$q = \frac{1}{\bar{h}} \tag{5.3}$$

where $\bar{h}$ is the average headway $(1/n \times \Sigma h_i)$. The importance of time headways in traffic analysis will be extensively demonstrated in forthcoming sections of this chapter.

The average traffic speed can be defined in two ways. The first is the arithmetic mean of speeds observed at some point. This is referred to as the time mean speed, $\bar{u}_t$, and is expressed as

$$\bar{u}_t = \frac{1}{n} \sum_{i=1}^{n} u_i \tag{5.4}$$

The second measure is more useful in the context of traffic analysis and is defined on the basis of the time necessary for a vehicle to traverse some known length of roadway, l. This measure is known as the space mean speed, u, and is

$$u = \frac{(1/n) \displaystyle\sum_{i=1}^{n} l_i}{\bar{t}} \tag{5.5}$$

Where l_i is the length of roadway used for the speed measurement of vehicle i, and

$$\bar{t} = \frac{1}{n} [t_1(l_1) + t_2(l_2) + \cdots + t_n(l_n)]$$

With $t_n(l_n)$ being the time necessary for vehicle n to traverse a section of roadway of length l. Note that if all vehicle speeds are measured over the same length of roadway $(L = l_1 = l_2 = \cdots l_n)$,

$$u = \frac{1}{(1/n) \displaystyle\sum_{i=1}^{n} [1/(L/t_i)]} \tag{5.6}$$

which is the harmonic mean of speed.

Finally, traffic density, k, refers to the number of vehicles occupying a length of roadway at a specified time. This is stated simply as

$$k = \frac{n}{l} \tag{5.7}$$

5.3 BASIC TRAFFIC STREAM MODELS

Based on the definitions presented in the preceding section, a simple identity provides the basic relationship between traffic flow, speed (space mean speed), and density,

$$q = uk \tag{5.8}$$

with typical units of flow (q), speed (u), and density (k) being vehicles per hour (veh/hr), miles per hour (mph) and vehicles per mile (veh/mi), respectively. As will be shown in the following discussion, Eq. 5.8 will serve the important function of linking specific models of traffic into a consistent generalized model.

5.3.1 Speed-Density Model

Perhaps the most intuitive starting point for developing the basic traffic model is to focus on the relationship between speed and density. Visualize a highway on which you are the only driver. Under these conditions the density (veh/mi) will be very low (near zero) and you will be able to travel freely at the posted speed limit. This is referred to as the free flow speed (denoted u_f), since vehicle operating speed is not inhibited by the presence of other vehicles. As more and more vehicles use the highway, the traffic density will increase and operating speeds will decline from the free-flow value, since vehicular maneuvers and cautious driving will tend to slow the traffic stream. Eventually, the highway will become so congested that the traffic will come to a stop ($u = 0$) and the density will be determined largely by the physical length of vehicles and the small spaces left between them. This high-density condition is referred to as the jam density, k_j.

One possible representation of the process described above might be the linear relationship indicated in Fig. 5.1, as suggested by Greenshield [Transportation Research Board Special Report 165, 1975]. Mathematically, this can be expressed as

$$u = u_f \left(1 - \frac{k}{k_j}\right) \tag{5.9}$$

The advantage of a linear representation of the speed-density relationship is that it will result in intuitive mathematical representations of traffic flow, speed, and density interactions. Various nonlinear forms of the speed–density relationship have been investigated by a number of researchers [Pipes 1967; Drew 1965], as a result of observed nonlinearities in the speed–density relationship near free-flow conditions and jam density (i.e., at the extremes of the relationship). However, for exposition, only the linear form will be presented here.

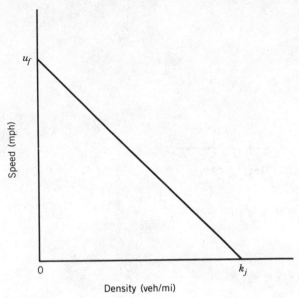

FIGURE 5.1
Illustration of a typical linear speed-density relationship.

5.3.2 Flow-Density Model

Based on the assumption of a linear relationship between speed and density, a parabolic flow-density model can be derived by substituting Eq. 5.9 into Eq. 5.8

$$q = u_f \left(k - \frac{k^2}{k_j} \right) \tag{5.10}$$

The general form of Eq. 5.10 is illustrated in Fig. 5.2. Two points on this figure are worthy of note. First the maximum flow rate, q_m, represents the highest rate of traffic flow that the highway is capable of supporting. This is also referred to as the traffic flow capacity or simply the capacity of the highway. The second point is the traffic density that corresponds to the maximum flow rate, k_m. There will also be a space mean speed, u_m, that will be the average speed of traffic at maximum flow. Equations for q_m, k_m, and u_m can be derived by differentiating Eq. 5.10. It is known that at maximum flow,

$$\frac{dq}{dk} = u_f \left(1 - \frac{2k}{k_j} \right)$$

$$= 0 \tag{5.11}$$

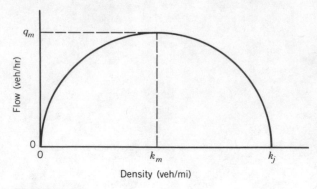

FIGURE 5.2
Illustration of the parabolic flow-density relationship.

and since the free flow speed (u_f) is not equal to zero,

$$k_m = \frac{k_j}{2}$$

(5.12)

Substituting Eq. 5.12 into Eq. 5.9 gives

$$u_m = u_f\left(1 - \frac{k_j}{2k_j}\right)$$

$$= \frac{u_f}{2}$$

(5.13)

and using Eq. 5.12 and Eq. 5.13 in Eq. 5.8 gives

$$q_m = u_m k_m$$

$$= \frac{u_f k_j}{4}$$

(5.14)

5.3.3 Speed-Flow Model

To develop a representation of the speed-flow relationship, Eq. 5.9 is rearranged to

$$k = k_j\left(1 - \frac{u}{u_f}\right)$$

(5.15)

and by substituting Eq. 5.15 into Eq. 5.8, we obtain

$$q = k_j\left(u - \frac{u^2}{u_f}\right)$$

(5.16)

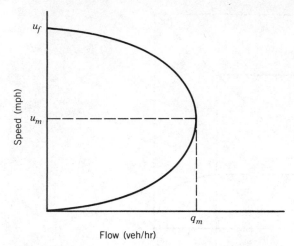

FIGURE 5.3
Illustration of the parabolic speed-flow relationship.

This again results in a parabolic function as shown in Fig. 5.3. All flow, speed, and density relationships and their interactions are graphically represented in Fig. 5.4.

Example 5.1

A section of highway has a free flow speed of 55 mph and a jam density of 367 vehicles per mile. Assuming a linear speed-density relationship, calculate the density at maximum flow, maximum flow rate, and speed at maximum flow.

Solution

Using Eq. 5.12, we find that the density at maximum flow is

$$k_m = \frac{k_j}{2}$$

$$= \frac{367}{2}$$

$$= \underline{\underline{183.5 \text{ veh/mi}}}$$

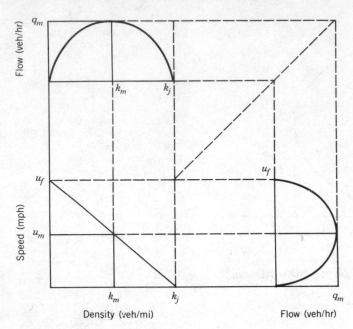

FIGURE 5.4
Flow-density, speed-density, and speed-flow relationships (assuming a linear speed-density model).

Upon using Eq. 5.14, the maximum flow rate is

$$q_m = \frac{u_f k_j}{4}$$

$$= \frac{55 \times 367}{4}$$

$$= 5046.25 \text{ veh/hr}$$

Finally, space mean speed at maximum flow is given by Eq. 5.13 as

$$u_m = \frac{u_f}{2}$$

$$= \frac{55}{2}$$

$$= 27.5 \text{ mph}$$

5.4 MODELS OF TRAFFIC FLOW

With the basic relationships between traffic flow, speed, and density formalized, attention can now be directed toward a more microscopic view of traffic flow. That is, instead of simply modeling the number of vehicles passing a point in some time interval, there is considerable analytic value in modeling the time between the arrivals of successive vehicles (i.e., the notion of vehicular headways presented earlier). The most simplistic approach to vehicle arrival modeling is to assume that all vehicles are equally or uniformly spaced. This results in what is termed a deterministic and uniform arrival pattern. Under this assumption, if the flow is 360 veh/hr the number of vehicles arriving in any 5-min time interval is 30 and the headway between all vehicles is 10 sec (since h will equal to $3600/q$). However, experience tells us that, in many instances, such uniformity of flow may not be an entirely realistic representation of traffic, since some 5-min intervals are likely to have more or less flow than other 5-min intervals. Thus a more elaborate model of vehicular arrivals is often warranted.

Models that account for the nonuniformity of flow are derived by assuming that the pattern of vehicle arrivals corresponds to some random process. The problem then becomes one of selecting a probability distribution that is a reasonable representation of observed traffic arrival patterns. An example of such a distribution is the Poisson distribution (the limitations of which will be discussed later), which is expressed as

$$P(n) = \frac{(\lambda t)^n e^{-\lambda t}}{n!} \qquad (5.17)$$

where t is the duration of the time interval over which vehicles are counted, $P(n)$ is the probability of having n vehicles arrive in time t, and λ is the average flow or arrival rate in vehicles per unit time.

Example 5.2

A roadway has an average hourly volume of 360 veh/hr. Assuming that the arrival of vehicles is Poisson distributed, estimate the probabilities of having 0, 1, 2, 3, 4, and 5 or more vehicles arriving over a 20-sec time interval.

Solution

The average arrival rate, λ, is 360 veh/hr or 0.1 vehicles per second (veh/sec). Using this in Eq. 5.17 with $t = 20$ sec, we find that the probabilities of having 0,

1, 2, 3, and 4 vehicles arrive are

$$P(0) = \frac{(0.1 \times 20)^0 e^{-0.1(20)}}{0!} = \underline{\underline{0.135}}$$

$$P(1) = \frac{(0.1 \times 20)^1 e^{-0.1(20)}}{1!} = \underline{\underline{0.271}}$$

$$P(2) = \frac{(0.1 \times 20)^2 e^{-0.1(20)}}{2!} = \underline{\underline{0.271}}$$

$$P(3) = \frac{(0.1 \times 20)^3 e^{-0.1(20)}}{3!} = \underline{\underline{0.180}}$$

$$P(4) = \frac{(0.1 \times 20)^4 e^{-0.1(20)}}{4!} = \underline{\underline{0.09}}$$

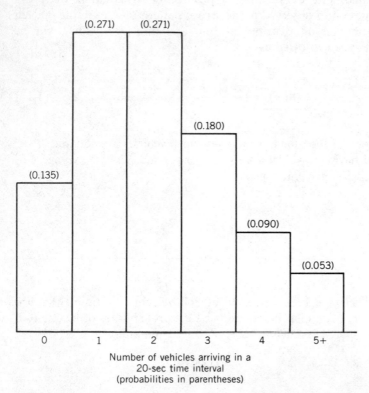

FIGURE 5.5
Histogram of the Poisson distribution for $\lambda = 0.1$ vehicles per second.

For five or more vehicles

$$P(5) = 1 - P(h < 5)$$
$$= 1 - 0.135 - 0.271 - 0.271 - 0.18 - 0.09$$
$$= \underline{\underline{0.053}}$$

A histogram of these probabilities is shown in Fig. 5.5.

The assumption of Poisson distributed vehicle arrivals also implies a distribution of the time intervals between the arrivals of successive vehicles (i.e., time headway). To show this, let the average arrival rate, λ, be in units of vehicles per second, so that

$$\lambda = \frac{q}{3600} \tag{5.18}$$

where q is the flow in vehicles per hour. Substituting Eq. 5.18 into 5.17 gives

$$P(n) = \frac{(qt/3600)^n e^{-qt/3600}}{n!} \tag{5.19}$$

Note that the probability of having no vehicles arrive in a time interval of length t (i.e., $P(0)$) is the equivalent of the probability of a vehicle headway, h, being greater than or equal to the time interval t. So from Eq. 5.19,

$$P(0) = P(h \ge t)$$
$$= e^{-qt/3600} \tag{5.20}$$

This distribution of vehicle headways is known as the negative exponential distribution and is often simply referred to as the exponential distribution.

Example 5.3

Again assuming that the arrival of vehicles is Poisson distributed with an hourly volume of 360 veh/hr (as in Example 5.2), determine the probability that the gap between successive vehicles will be less than 8 sec.

Solution

By definition, $P(h < t) = 1 - P(h \geq t)$. Therefore, using Eq. 5.20 yields

$$P(h < t) = 1 - e^{-qt/3600}$$
$$= 1 - e^{-360(8)/3600}$$
$$= \underline{\underline{0.551}}$$

To help visualize the shape of the exponential distribution, Fig. 5.6 shows the probability distribution implied by Eq. 5.20 with the flow, q, equal to 360 veh/hr as in Example 5.3.

Empirical observations have shown that the assumption of Poisson-distributed traffic arrivals is most realistic in lightly congested traffic conditions. As traffic flows become heavily congested or when traffic signals cause cyclical traffic stream disturbances, other distributions of traffic flow become more appropriate.

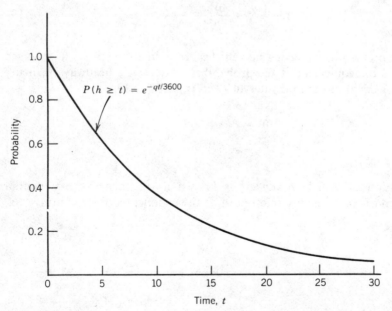

FIGURE 5.6
Exponentially distributed probabilities of headways greater than or equal to t, with $q = 360$ veh/hr.

Such distributions are discussed in detail in more specialized sources [Transportation Research Board Special Report 165, 1975; Haight 1965].

5.5 QUEUING THEORY AND TRAFFIC FLOW ANALYSIS

The formation of traffic queues at intersections and roadway bottlenecks during congested periods is a source of considerable time delay and results in the loss of highway performance. Under some extreme conditions queuing delay can account for 90 percent or more of a motorist's total trip travel time. Thus it is clear that being able to understand the concept of queuing and to develop appropriate mathematical models is an important need in traffic analysis.

As is well known, the problem of queuing is not unique to highway traffic analysis. Other nontransportation fields, such as the design and operation of industrial plants, retail stores, and service-oriented industries, must give serious attention to the problem of queuing. The impact that queues have on performance and productivity in manufacturing and other fields has led to numerous theories of queuing behavior (i.e., the process by which queues form and dissipate). As will be shown, the models of traffic flow presented in the previous section (i.e., deterministic and Poisson arrivals) will form the basis for studying traffic queues within the more general context of queuing theory.

5.5.1 Dimensions of Queuing Models

The purpose of traffic queuing models is to provide a means to estimate important measures of highway performance including vehicular delay and traffic queue lengths. Such estimates are critical to highway design (e.g., length of turning bays and number of lanes at intersections) and traffic operations control including the timing of traffic signals at intersections.

Queuing models are derived from underlying assumptions with regard to arrival patterns, departure characteristics, and queue disciplines. Section 5.4 explored traffic arrival patterns where, given an average vehicle arrival rate, two possible distributions of the time between the arrival of successive vehicles were considered: (1) equal time intervals (derived from the assumption of uniform or deterministic arrivals) and (2) exponentially distributed time intervals (derived from the assumption of Poisson arrivals). In addition to vehicle arrival assumptions, the development of traffic queuing models requires assumptions relating to vehicle departure characteristics. Of particular interest is the distribution of the amount of time it takes a vehicle to depart, for example, the time to pass through an intersection at the beginning of a green signal or the time to decide to enter traffic from a stop sign. Again, given an average vehicle departure rate, the assumption of a deterministic or exponential distribution of departure times is appropriate. Another important aspect of queuing models is the number of available departure channels. For most traffic applications only one departure channel will exist, such as a lane or group of lanes passing through an intersection. However, multiple departure channels are encountered in some traffic applications such as at toll booths at entrances to turnpikes or bridges.

The final necessary assumption relates to the queue discipline. Basically, two options have been popularized in the development of queuing models: first-in-first-out (FIFO), indicating that the first vehicle to arrive is the first to depart, and last-in-first-out (LIFO), indicating that the last vehicle to arrive is the first to depart. For virtually all traffic-oriented queues, the FIFO queuing discipline is the more realistic of the two.

Specific arrival and departure assumptions define a queuing regime. Queuing regimes are identified by three alphanumeric values: the first value indicates the arrival rate assumption; the second gives the departure rate assumption; and the third value indicates the number of departure channels. For traffic arrival and departure rate assumptions, the deterministic or uniform distribution is denoted D and the exponential distribution is denoted M. Thus a $D/D/1$ queuing regime assumes deterministic arrivals and departures with one departure channel. Similarly, an $M/D/1$ queuing regime assumes exponentially distributed arrival times, deterministic (uniform) departure times, and one service channel.

5.5.2 $D/D/1$ Queuing Regime

The concept of queuing models is most readily understood by considering the case of deterministic arrival and departure rates with a single departure channel. This queuing regime lends itself to an intuitive graphical or mathematical solution that is best illustrated by example.

Example 5.4

Vehicles arrive at an entrance to a national park. There is a single gate (at which all vehicles must stop), where a ranger distributes a free brochure. The park opens at 8:00 a.m., at which time vehicles begin to arrive at the rate of 480 veh/hr. After 20 min, the flow rate declines to 120 veh/hr and continues at that level for the remainder of the day. If the time required to distribute the brochure is 15 sec, describe the operational characteristics of a $D/D/1$ regime.

Solution

First arrival and departure rates are put in common units of vehicles per minute.

$$\lambda = \frac{480 \text{ veh/hr}}{60} = 8 \text{ veh/min} \qquad \text{for } t \le 20$$

$$\lambda = \frac{120 \text{ veh/hr}}{60} = 2 \text{ veh/min} \qquad \text{for } t > 20$$

$$\mu = \frac{60 \text{ sec/min}}{15 \text{ sec/veh}} = 4 \text{ veh/min} \qquad \text{for all } t$$

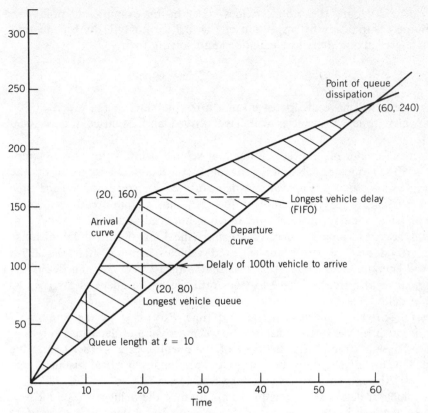

FIGURE 5.7
$D/D/1$ queuing diagram for Example 5.4.

More generally, equations for the total number of vehicles that have arrived and departed up to a specified time, t, can be written. Define t as the number of minutes after the start of the queuing process. The total number of vehicle arrivals at time t is equal to

$$8t \quad \text{for } t \le 20$$

and

$$160 + 2(t - 20) \quad \text{for } t > 20$$

Similarly, the number of vehicle departures is

$$4t \quad \text{for all } t$$

The above equations can be illustrated graphically as shown in Fig. 5.7. When the arrival curve is above the departure curve, a queuing condition will exist. The point at which the arrival curve falls below the departure curve is the moment that the queue dissipates (i.e., no more lines exist). In this example, the point can be determined graphically by inspection of Fig. 5.7, or analytically by equating appropriate arrival and departure equations and solving for t,

$$160 + 2(t - 20) = 4 \quad \text{or} \quad t = 60 \text{ min}$$

Thus the queue will have dissipated 60 min after the start of the process (9:00 a.m.) at which time 240 vehicles will have arrived and departed (i.e., 4×60 vehicles).

Another parameter of interest is individual vehicle delay. Under the assumption of a FIFO queuing discipline, the delay of a vehicle is given by the horizontal distance between arrival and departure curves starting from the time of vehicle arrival. So, by inspection of Fig. 5.7, the one-hundredth vehicle to arrive will have a delay of 12.5 min, the one-hundred-and-sixtieth vehicle to arrive will have the longest delay of 20 min (the longest horizontal distance between arrival and departure curves), and vehicles arriving after the 239th vehicle will encounter no delay. It follows that with a LIFO queuing discipline, the first vehicle to arrive will have to wait until the entire queue clears (i.e., 60 minutes of delay).

The total length of queue at a specified time, expressed by the number of vehicles, is given by the vertical distance between arrival and departure curves at that time. Thus at 10 min after the start of the queuing process (8:10 a.m.) the queue is 40 vehicles long, and the longest queue (longest vertical distance) will occur at $t = 20$ min and is 80 vehicles long (see Fig. 5.7).

Total vehicular delay, defined as the summation of the delays of each individual vehicle, is given by the total area between arrival and departure curves (see Fig. 5.7) and is in units of vehicle-minutes. In this example, the areas between arrival and departure curves can be determined by summing triangular areas, giving total delay as

$$\text{total delay} = \tfrac{1}{2}(80 \times 20) + \tfrac{1}{2}(80 \times 40)$$
$$= 2400 \text{ veh-min}$$

and, finally, since 240 vehicles encounter delay (as previously determined), the average delay per vehicle is 10 min (2400 veh-min/240 veh).

Example 5.5

After observing arrivals and departures at a toll both over a 60-min time period, a transportation student notes that the arrival and service rates are deterministic but, instead of being uniform, they change over time according to a known function. The arrival rate is given by the function $\lambda(t) = 2.2 + 0.17t - 0.0032t^2$ and the departure rate is given by $\mu(t) = 1.2 + 0.07t$, where t is in minutes after the beginning of the observation period. Determine the total vehicular delay at the toll both and longest queue (assume $D/D/1$ queuing).

Solution

The time-to-queue dissipation can be determined by solving,

$$\int_0^t 2.2 + 0.17t - 0.0032t^2 \, dt = \int_0^t 1.2 + 0.07t \, dt$$

$$2.2t + 0.085t^2 - 0.00107t^3 = 1.2t + 0.035t^2$$

$$- 0.00107t^3 + 0.05t^2 + t = 0$$

which gives $t = 61.8$ min. Therefore, the total vehicular delay (which is the area between arrival and departure functions) is given simply as,

$$D = \int_0^{61.8} 2.2t + 0.085t^2 - 0.00107t^3 \, dt - \int_0^{61.8} 1.2t + 0.035t^2$$

$$= 1.1t^2 + 0.0283t^3 - 0.0002675t^4 - 0.6t^2 - 0.0117t^3 \, \big|_0^{61.8}$$

$$= -0.0002675(61.8)^4 + 0.0166(61.8)^3 + 0.5(61.8)^2$$

$$= \underline{1925.8 \text{ veh-min}}$$

The queue length at any time t is given by the function,

$$Q(t) = \int_0^t 2.2 + 0.17t - 0.0032t^2 \, dt - \int_0^t 1.2 + 0.07t \, dt$$

$$= -0.00107t^3 + 0.05t^2 + t$$

Solving for the maximum queue length gives

$$\frac{dQ(t)}{dt} = -0.00321t^2 + 0.1t + 1$$

$$= 0$$

$$t = \underline{39.12 \text{ min}}$$

and, by substitution with $t = 39.12$ min,

$$Q(39.12) = -0.00107t^3 + 0.05t^2 + t \, \big|_0^{39.12}$$

$$= -0.00107(39.12)^3 + 0.05(39.12)^2 + 39.12$$

Therefore,

$$\underline{Q_m = 51.58 \text{ veh}}$$

This problem is an example of a time-dependent deterministic queue, since the deterministic arrival and departure rates change over time.

5.5.3 $M/D/1$ Queuing Regime

The assumption of exponentially distributed times between the arrivals of successive vehicles will, in some cases, result in a more realistic representation of traffic flow than the assumption of uniformly distributed arrival times (as discussed in Section 5.4). Therefore, the $M/D/1$ queuing regime has some important applications within the traffic analysis field. Although a graphical solution to an $M/D/1$ queue is difficult, a mathematical solution is quite straightforward. Defining a traffic intensity term, ρ, as the ratio of average arrival to departure rates (λ/μ), and assuming that ρ is less than 1, it can be shown that for an $M/D/1$ queue the averge length of queue is [Gelenbe and Pujolle 1987].

$$\overline{Q} = \frac{2\rho - \rho^2}{2(1 - \rho)} \tag{5.21}$$

the average waiting time in the queue is

$$\overline{w} = \frac{\rho}{2\mu(1 - \rho)} \tag{5.22}$$

and the average time spent in the system (i.e., the summation of average queue waiting time and the average departure time) is

$$\overline{t} = \frac{2 - \rho}{2\mu(1 - \rho)} \tag{5.23}$$

It is important to recognize in passing that under the assumption that traffic intensity is less than unity (i.e., $\lambda < \mu$) the $D/D/1$ regime will predict no queue formation. However, queuing regimes with random arrivals or departures, such as the $M/D/1$, will predict queue formations under such conditions. Also, note that the $M/D/1$ queuing regime presented here is based on steady-state conditions (i.e., constant average arrival and departure rates) with randomness given from an assumed probability distribution. This contrasts with the time-varying deterministic queuing case, as presented in Example 5.5, in which arrival and departure rates changed over time, but randomness was not present.

Example 5.6

Consider the national park described in Example 5.4. However, let the average arrival flow rate be 180 veh/hr over the entire period from park opening time (8:00 a.m.) until closing at dusk. Describe $M/D/1$ operational characteristics.

Solution

Putting arrival and departure rates into common units of vehicles per minute gives

$$\lambda = \frac{180 \text{ veh/hr}}{60}$$
$$= 3 \text{ veh/min}$$
$$\mu = 4 \text{ veh/min}, \quad \text{as before}$$

and

$$\rho = \frac{\lambda}{\mu}$$
$$= 0.75$$

For the average length of queue, Eq. 5.21 is applied,

$$\overline{Q} = \frac{2(0.75) - 0.75^2}{2(1 - 0.75)}$$
$$= 1.87 \text{ veh}$$

For the average waiting time in the queue, Eq. 5.22 gives

$$\overline{w} = \frac{0.75}{2(4)(1 - 0.75)}$$
$$= 0.375 \text{ min}$$

For average time spent in the system, Eq. 5.23 is used,

$$\overline{t} = \frac{2 - 0.75}{2(4)(1 - 0.75)}$$
$$= 0.625 \text{ min}$$

or

$$\overline{t} = \overline{w} + \text{average service time} \left(\frac{1}{\mu} \right)$$
$$= 0.375 + 0.25$$
$$= 0.625 \text{ min}$$

5.5.4 $M/M/1$ Queuing Regime

A queuing regime that assumes exponentially distributed departure time patterns in addition to exponentially distributed arrival times is also useful in some traffic applications. For example, exponentially distributed departure pattens might be a reasonable assumption at a toll booth where some arriving drivers have the correct toll and can be processed quickly, and others may not have the correct toll, thus producing a distribution of departures about some mean departure rate. Under standard $M/M/1$ assumptions, it can be shown that the average length of queue is (again assuming that ρ is less than 1),

$$\bar{Q} = \frac{\rho^2}{1 - \rho} \qquad (5.24)$$

the average waiting time in the queue is

$$\bar{w} = \frac{\lambda}{\mu(\mu - \lambda)} \qquad (5.25)$$

and the average time spent in the system is

$$\bar{t} = \frac{1}{\mu - \lambda} \qquad (5.26)$$

For further description of the operational characteristics of these and other queuing regimes, the reader is referred to other sources [Gelenbe and Pujolle 1987; Haight 1963].

Example 5.7

The park ranger (see Examples 5.4 and 5.6) takes an average of 15 sec to distribute brochures, but the time varies depending on whether or not patrons have questions relating to park operating policies. Given the conditions of Example 5.6 (i.e., an average arrival rate of 180 veh/hr) describe $M/M/1$ operational characteristics.

Solution

With average arrival and departure rates and traffic intensity as determined in Example 5.6, Eq. 5.24 gives the average length of queue as

$$\bar{Q} = \frac{0.75^2}{1 - 0.75}$$
$$= 2.25 \text{ veh}$$

The average waiting time in the queue is (from Eq. 5.25)

$$\bar{w} = \frac{3}{4(4 - 3)}$$
$$= 0.75 \text{ min}$$

The average waiting time in the system is (from Eq. 5.26)

$$\bar{t} = \frac{1}{4 - 3}$$

$$= \underline{\underline{1 \text{ min}}}$$

5.6 TRAFFIC ANALYSIS AT HIGHWAY BOTTLENECKS

Some of the most severe congestion problems occur at highway bottlenecks, which can be generally defined as a portion of highway with lower capacity (q_m) than the incoming section of roadway. This reduction is capacity can originate from a variety of sources such as an effective decrease in the number of through traffic lanes, reduced shoulder widths, and the presence of traffic signals (details of roadway capacity determination will be presented in Chapter 6). Two types of traffic bottlenecks can be identified: (1) recurring bottlenecks and (2) incident-provoked bottlenecks. Recurring bottlenecks occur where the highway itself limits capacity by, for example, a physical reduction in the number of lanes. Such bottlenecks result from typical recurring traffic flows that exceed the restrictive vehicular capacity of the bottleneck area. In contrast, incident-provoked bottlenecks occur as a result of vehicle breakdowns or accidents that effectively reduce highway capacity by restricting the through movement of traffic. Since incident-provoked bottlenecks are unanticipated and temporary in nature, they have features that distinguish them from recurring bottlenecks such as the fact that incident bottlenecks often have capacities that vary over time. For example, an accident may initially stop traffic flow completely, but as the wreckage is cleared, partial capacity (e.g., one lane open) may be provided for a period of time before full capacity is eventually restored. A feature shared by both recurring and incident-provoked bottlenecks is the adjustment in traffic flow that may occur as travelers choose other routes, to avoid the bottleneck area, in response to visual information or traffic advisories.

With a basic understanding of the concept of traffic bottlenecks, an analytic assessment of traffic flow, under restrictive capacity conditions, can be undertaken with the traffic queuing models discussed in Section 5.5. A particularly intuitive approach from which traffic congestion at bottlenecks can be analyzed is the assumption of a $D/D/1$ queuing regime.

Example 5.8

A freeway has a direction capacity of 4000 veh/hr and a constant flow of 2900 veh/hr during the morning commute to work (i.e., no adjustments to traffic flow are produced by the incident). At 8:00 a.m. a traffic accident closes the freeway to all flow. At 8:12 a.m. the freeway is partially opened with a capacity of 2000 veh/hr. Finally, the wreckage is removed and the freeway is restored to full capacity (4000 veh/hr) at 8:31 a.m. Assume a $D/D/1$ queuing regime determine total delay, longest queue length, time of queue dissipation, and longest wait (assuming FIFO).

Solution

Let μ be the full capacity departure rate and μ_r be the restrictive partial capacity departure rate. Putting arrival and departure rates in common units of vehicles per minute,

$$\mu = \frac{4000}{60}$$
$$= 66.67 \text{ veh/min}$$
$$\mu_r = \frac{2000}{60}$$
$$= 33.33 \text{ veh/min}$$
$$\lambda = \frac{2900}{60}$$
$$= 48.33 \text{ veh/min}$$

The total number of arriving vehicles is constant over the entire time period and equal to λt, where t is the number of minutes after 8:00 a.m. The total number of departing vehicles is

$$\begin{array}{ll} 0 & \text{for } t \le 12 \\ \mu_r(t - 12) & \text{for } 12 < t \le 31 \\ 633.3 + \mu(t - 31) & \text{for } t > 31 \end{array}$$

The above arrival and departure rates can be represented graphically, as shown in Fig. 5.8. As discussed in Section 5.5, for the $D/D/1$ queuing regime, the queue will dissipate at the intersection point of arrival and departure curves, which can be determined as

$$\lambda t = 633.3 + \mu(t - 31) \quad \text{or} \quad t = 78.16 \text{ min} \quad \text{(just after 9:18 a.m.)}$$

At this time a total of 3777.5 vehicles (48.33×78.16) will have arrived and departed (for the sake of clarity, fractions of vehicles are used). The longest

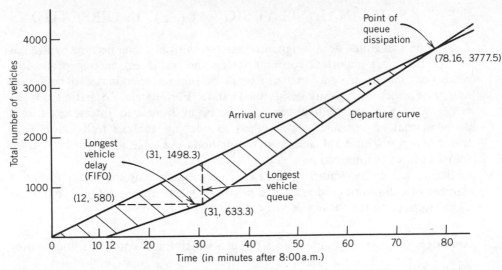

FIGURE 5.8
$D/D/1$ queuing diagram for Example 5.8.

queue (longest vertical distance between arrival and departure curves) occurs at
8:31 a.m. and is

$$Q_m = \lambda t - \mu_r(t - 12)$$

$$= 48.33(31) - 33.33(19)$$

$$= \underline{\underline{865 \text{ veh}}}$$

Total vehicular delay is (using equations for triangular and trapezoidal areas to
calculate the total area between arrival and departure curves)

$$D = \tfrac{1}{2}(12)(580) + \tfrac{1}{2}(580 + 1498.3)(19) - \tfrac{1}{2}(19)(633.3)$$

$$+ \tfrac{1}{2}(1498.3 - 633.3)(78.16 - 31)$$

$$= \underline{\underline{37{,}604.2 \text{ veh-min}}}$$

The average delay per vehicle is 9.95 min (37604.2/3777.5). The longest wait of
any vehicle (the longest horizontal distance between arrival and departure curves)
assuming a FIFO queuing discipline, will be the delay time of the 633.3 vehicle to
arrive. This vehicle will arrive 13.1 min (633.3/48.3) after 8:00 a.m. and will
depart at 8:31 a.m., thus being delayed a total of 17.9 min.

5.7 TRAFFIC ANALYSIS AT SIGNALIZED INTERSECTIONS

The analysis of traffic flow at signalized intersections has long been recognized as one of the most important concerns facing the traffic engineering profession, since the amount of delay that can occur at such intersections can render an otherwise excellent highway design inadequate. Fortunately, estimating and developing means of avoiding vehicular delay at signalized intersections is a problem that is particularly well suited to queuing analysis techniques. Thus traffic engineers have valuable analytic methods available to arrive at efficient signal designs at intersections.

Before presenting traffic signal analysis principles, it is important to give a number of definitions (and associated notations where appropriate) of commonly used intersection related terminology. These terms include the following:

Approach   A lane or group of lanes through which traffic enters the intersection.

Cycle   One complete sequence (for all approaches) of signal indications (greens, yellows, reds, etc.).

Cycle length   The total time for the signal to complete one cycle (given the symbol c).

Interval   A period of time during which all signal indications remain constant.

Change Interval   The yellow plus all red intervals that provide for clearance of the intersection before conflicting traffic movements are released (expressed in seconds).

Green Time   The time within a cycle in which an approach has the green indication (stated in seconds).

Lost Time   Time during which the intersection is not effectively used by any approach. These times occur during the change interval (when the intersection is cleared), and at the beginning of each green indication as the first few cars in a standing queue experience start-up delays.

Effective Green   The time that is effectively available to the permitted traffic movements. This is generally taken to be the green time plus the change interval minus the lost time for the approach, stated in seconds and given the symbol g.

Effective Red   The time during which a given traffic movement is effectively not permitted to move. Stated in seconds, it is the cycle length minus the effective green time (given the symbol r).

Saturation Flow   The maximum flow that could pass through an intersection, from a given approach if that approach were allocated all of the cycle time as effective green with no lost time (given the symbol s).

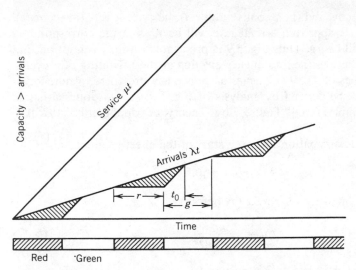

FIGURE 5.9
$D/D/1$ intersection queuing with approach capacity exceeding arrivals.

Approach Capacity The maximum flow that can pass through an intersection under prevailing roadway and traffic conditions, given the effective green time allocated to the approach. It is the saturation flow times the ratio of effective green to cycle length ($C = s \times g/c$).

Signal Timing The operating characteristics of the signal with the parameters being signal cycle length and the effective green and effective red times allocated to intersection approaches.

Pretimed Signal A signal whose signal timing (i.e., cycle length and allocation of effective greens and reds to intersection approaches) is fixed over specified time periods and does not change in response to chances in traffic flow at the intersection.

5.7.1 Intersection Analysis with a $D/D/1$ Queuing Regime

As was the case in the analysis of highway bottlenecks, the assumption of a $D/D/1$ queuing regime gives a strong intuitive appeal to signalized intersection analysis. To illustrate this, consider the case where approach capacity exceeds approach arrivals. Under these conditions, the assumption of a $D/D/1$ queuing regime will result in a queuing system as shown in Fig. 5.9, where λ is the arrival rate (typically in vehicles per second), μ is the departure rate (in vehicles per second), t is the total transpired time (in seconds), t_0 is the time after the start of effective green until queue dissipation (in seconds), g is the effective green (in seconds), r is the effective red (in seconds), and c is the cycle length (in seconds). Note that the per-cycle approach arrival rate will be λc and the corresponding approach capacity will be μg. Thus Fig. 5.9 is predicated on the assumption that μg exceeds λc for all cycles (i.e., no queues existing at the beginning of a cycle).

Given the properties of $D/D/1$ queues as previously presented, a number of general equations can be derived by analysis of Fig. 5.9, as was done earlier in $D/D/1$ queuing examples (see ["Traffic Flow Theory: A Monograph," 1975]).

1. The time to queue dissipation after the start of the effective green, t_0,

$$\lambda(r + t_0) = \mu t_0$$

and if the traffic intensity, $\rho = \lambda/\mu$ (as before),

$$t_0 = \frac{\rho r}{(1 - \rho)} \tag{5.27}$$

2. The proportion of the cycle with a queue, P_q,

$$P_q = \frac{(r + t_0)}{c} \tag{5.28}$$

3. The proportion of vehicles stopped, P_s,

$$P_s = \frac{\lambda(r + t_0)}{\lambda(r + g)}$$

$$= \frac{t_0}{\rho c} \tag{5.29}$$

4. The maximum number of vehicles in the queue, Q_m,

$$Q_m = \lambda r \tag{5.30}$$

5. The total vehicular delay per cycle, D,

$$D = \frac{\lambda r^2}{2(1 - \rho)} \tag{5.31}$$

6. The average vehicular delay per cycle, d,

$$d = \frac{\lambda r^2}{2(1 - \rho)} \times \frac{1}{\lambda c}$$

$$= \frac{r^2}{2c(1 - \rho)} \tag{5.32}$$

7. The maximum delay of any vehicle (assuming a FIFO queuing discipline), d_m,

$$d_m = r \tag{5.33}$$

Example 5.9

An approach at a pretimed signalized intersection has a saturation flow of 2400 veh/hr and is allocated 24 sec of effective green in an 80-sec signal cycle. If the flow at the approach is 500 veh/hr, provide an analysis of the intersection assuming a $D/D/1$ regime.

Solution

Putting arrival and departure rates in common units of vehicles per second,

$$\mu = \frac{2400}{3600}$$
$$= 0.67 \text{ veh/sec}$$
$$\lambda = \frac{500}{3600}$$
$$= 0.139 \text{ veh/sec}$$

Which gives a traffic intensity,

$$\rho = \frac{0.139}{0.67}$$
$$= 0.207$$

Checking to make certain that capacity exceeds arrivals, we obtain

$$\mu g = 0.67(24)$$
$$= 16.08 \text{ veh}$$
$$> \lambda c = 0.139(80)$$
$$= 11.12 \text{ veh}$$

Therefore, Eqs. 5.27 to 5.33 are valid. By definition, $r = c - g = 80 - 24 = 56$ sec. Thus for,
(a) Time to queue dissipation (Eq. 5.27),

$$t_0 = \frac{0.207(56)}{1 - 0.207}$$
$$= \underline{14.62 \text{ sec}}$$

(b) Proportion of the cycle with a queue (Eq. 5.28),

$$P_q = \frac{56 + 14.62}{80}$$
$$= \underline{\underline{0.883}}$$

(c) Proportion of vehicles stopped (Eq. 5.29),

$$P_s = \frac{14.62}{0.207(80)}$$

$$= 0.883$$

(d) Maximum number of vehicles in the queue (Eq. 5.30),

$$Q_m = 0.137(56)$$

$$= 7.78 \text{ veh}$$

(e) Total delay per cycle (Eq. 5.31),

$$D = \frac{0.139(56)^2}{2(1 - 0.207)}$$

$$= 274.84 \text{ veh-sec}$$

(f) Average delay per vehicle (Eq. 5.32),

$$d = \frac{56^2}{2(80)(1 - 0.207)}$$

$$= 24.72 \text{ sec/veh}$$

(g) Maximum delay of any vehicle (Eq. 5.33),

$$d_m = r$$

$$= 56 \text{ sec}$$

As previously stated, Eqs. 5.27 through 5.33 are valid only when the approach capacity exceeds arrivals. For the case where approach arrivals exceed capacity for some signal cycles, basic $D/D/1$ principles must again be used as demonstrated below.

Example 5.10

An approach to an intersection has a saturation flow of 1700 veh/hr. The signal's pretimed cycle length is 60 sec and the effective red is 40 sec. During three consecutive cycles 15, 8, and 4 vehicles arrive. Determine the total vehicular delay over the three cycles assuming a $D/D/1$ queuing regime.

Solution

For all cycles, the departure rate is

$$\mu = \frac{1700}{3600}$$

$$= 0.472 \text{ veh/sec}$$

During the first cycle, the number of vehicles that will depart from the signal is (permitting fractions of vehicles for the sake of clarity),

$$\mu g = 0.472(20)$$

$$= 9.44 \text{ veh}$$

Therefore, 5.56 vehicles (15 − 9.44) will not be able to pass through the intersection on the first cycle even though they arrive during the first cycle. At the end of the second cycle, 23 vehicles (15 + 8) will have arrived but only 18.88 ($2\mu g$) will have departed, leaving 4.12 vehicles waiting at the beginning of the third cycle. At the end of the third cycle, a total of 27 vehicles will have arrived and as many as 28.32 ($3\mu g$) could have departed so the queue will have dissipated at some time during the third cycle. This process is shown graphically in Fig. 5.10. From this figure, the total vehicular delay of the first cycle is (the area between arrival and departure curves),

$$D_1 = \tfrac{1}{2}(60)(15) - \tfrac{1}{2}(20)(9.44)$$

$$= 355.6 \text{ veh-sec}$$

Similarly, the delay in the second cycle is

$$D_2 = \tfrac{1}{2}(60)(15 + 23) - (40)(9.44) - \tfrac{1}{2}(20)(9.44 + 14.88)$$

$$= 479.2 \text{ veh-sec}$$

To determine the delay of the third cycle, it is necessary to first know exactly when, in this cycle, the queue dissipates. The time to queue dissipation after the start of the effective green, t_0, is (with λ_3 being the arrival rate during the third cycle),

$$\text{vehicles waiting at start of cycle} + \lambda_3(r + t_0) = \mu t_0$$

where

$$\lambda_3 = \frac{4}{60}$$

$$= 0.067 \text{ veh/sec}$$

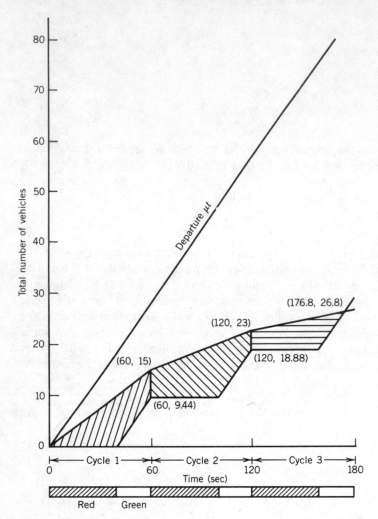

FIGURE 5.10
$D/D/1$ queuing diagram for Example 5.10.

Therefore,

$$4.12 + 0.067(40 + t_0) = 0.472t_0$$

or

$$\underline{\underline{t_0 = 16.8 \text{ sec}}}$$

Thus the queue will clear 56.8 sec (40 + 16.8) after the start of the third cycle; at which time a total of 26.8 vehicles ($0.067 \times 56.8 + 15 + 8$) will have arrived and departed from the intersection. The vehicular delay for the third cycle is

$$D_3 = \tfrac{1}{2}(56.8)(23 + 26.8) - 40(18.88) - \tfrac{1}{2}(16.8)(18.88 + 26.88)$$
$$= 275.41 \text{ veh-sec}$$

giving the total delay over all three cycles as

$$D = D_1 + D_2 + D_3$$
$$= 1110.2 \text{ veh-sec} \quad \text{or} \quad 18.5 \text{ veh-min}$$

5.7.2 Intersection Analysis with Probabilistic Arrivals

The assumption of nonuniform traffic arrivals, although often more realistic than the uniform (deterministic) arrival assumption, adds considerable complexity to the analysis of delay at signalized intersections. Over the past few decades, a number of delay formulas have been derived to account for the randomness of vehicle arrivals at signalized intersections. Among the better known of these delay formulas are those developed by Webster [1958] and Allsop [1972]. The Webster formula for approach delay at a pretimed signalized intersection is

$$d' = d + \frac{x^2}{2\lambda(1-x)} - 0.65\left(\frac{c}{\lambda^2}\right)^{1/3} x^{2+5\lambda} \tag{5.34}$$

where d' is the average vehicle delay, d is the average vehicle delay computed by assuming a $D/D/1$ queuing regime as shown in Eq. 5.32, x is the ratio of approach arrivals to approach capacity (i.e., $\lambda c/\mu g$), c is the cycle length, and λ is the average vehicle arrival rate. The Webster formula was developed by using data from a computer simulation of intersection operations. Allsop noted that the second term in Webster's equation is obtained by assuming an additional queue interposed between the arriving traffic and the signal, and the third term is merely an empirical correction ranging from 5 to 15 percent of total mean delay. Based on these observations, Allsop suggested that average delay, with random vehicular arrivals, be computed as

$$d' = \frac{9}{10}\left[d + \frac{x^2}{2\lambda(1-x)}\right] \tag{5.35}$$

For information on other intersection delay formulas that are based on the assumption of random vehicle arrivals, the reader is referred to other sources ["Traffic Flow Theory: A Monograph" 1975; *Highway Capacity Manual* 1985].

Example 5.11

Compute the average approach delay per cycle using Webster's and Allsop's formulas given the conditions described in Example 5.9.

Solution

As computed in Example 5.9, the delay assuming uniform arrivals, d, is

$$d = 24.72 \text{ sec/veh}$$

The ratio of approach arrivals to approach capacity is

$$x = \frac{\lambda c}{\mu g}$$

$$= \frac{0.139(80)}{0.67(24)}$$

$$= 0.69$$

Using Webster's formula (Eq. 5.34), we find that

$$d' = 24.72 + \frac{0.69^2}{2(0.139)(1 - 0.69)} - 0.65\left(\frac{80}{0.139^2}\right)^{1/3} 0.69^{2 + 5(0.139)}$$

$$= 24.72 + 5.52 - 3.84$$

$$= \underline{\underline{26.4 \text{ sec/veh}}}$$

Using Allsop's formula (Eq. 5.35), we obtain

$$d' = \frac{9}{10}\left[24.72 + \frac{0.69^2}{2(0.139)(1 - 0.69)}\right]$$

$$= 27.22 \text{ sec/veh}$$

5.7.3 Optimal Traffic Signal Timing

With a basic understanding of traffic flow at signalized intersections, it is natural to attempt to allocate available effective green times to competing approaches (i.e., competing for the right-of-way) in some optimal fashion. Although the traffic engineering profession has struggled with this concept for as long as signalized intersections have been in existence, a drive on many arterials suggests that optimal signal timing has not yet been universally achieved. The problem of optimal timing is complicated by a number of concerns. For example, although the distribution of traffic can be approximated by some arrival assumptions, to properly optimize timing, the arrival pattern of vehicles must be known with certainty. Traffic sensors embedded near intersections have resulted in traffic signal systems that respond to variations in traffic flow and thus have gone a long way to satisfy the traffic arrival pattern information need, but more fundamentally, there is a larger question that pervades signal optimization. Specifically, on what basis should traffic signals be optimized? Should total vehicular delay be minimized or should the total number of vehicle stops be minimized? Since these two minimization objectives usually provide different results, many signal timing routines seek to minimize a combined function that has both stops and delays as arguments [Vincent, Mitchell, and Robertson 1980]. More recent theoretical work has departed from the delay/stop approach and has demonstrated optimal signal timings based on the maximization of travelers' economic welfare [Mannering 1988]. Needless to say, the optimization measurement problem makes this aspect of traffic analysis a fruitful area for future research.

To demonstrate one possible signal optimization strategy (putting the optimization issues discussed above aside), assume that the sole objective of signal timing is to minimize total vehicular delay. Such a strategy is illustrated by the following example.

Example 5.12

A pretimed signal controls a four-way intersection with no turning permitted and zero lost time. The eastbound (eb) and westbound (wb) traffic volumes are 700 and 800 veh/hr, respectively, and both movements share the same red and green portions of the cycle. The northbound (nb) and southbound (sb) directions also share cycle times with volumes of 400 and 250 veh/hr, respectively. If the saturation flow of all approaches is 1800 veh/hr, the cycle length is 60 sec and a $D/D/1$ queuing regime holds, determine the effective green and red times that must be allocated to each directional combination (i.e., north–south, east–west) to minimize total vehicular delay.

Solution

Putting arrival and departure rates in common units of vehicles per second, we obtain

$$\lambda_{eb} = \frac{700}{3600} = 0.194 \text{ veh/sec}$$

$$\lambda_{wb} = \frac{800}{3600} = 0.222 \text{ veh/sec}$$

$$\lambda_{nb} = \frac{400}{3600} = 0.111 \text{ veh/sec}$$

$$\lambda_{sb} = \frac{250}{3600} = 0.069 \text{ veh/sec}$$

$$\mu = \frac{1800}{3600} = 0.5 \text{ veh/sec}$$

Since the departure rate is the same for all approaches, the traffic intensities are

$$\rho_{eb} = 0.388$$
$$\rho_{wb} = 0.444$$
$$\rho_{nb} = 0.222$$
$$\rho_{sb} = 0.138$$

Assume that capacity exceeds arrivals for all approaches. Thus, by summing the total delays of each approach (using Eq. 5.31), the total vehicular delay at the intersection is

$$D = \frac{\lambda_{eb} r_{eb}^2}{2(1 - \rho_{eb})} + \frac{\lambda_{wb} r_{wb}^2}{2(1 - \rho_{wb})} + \frac{\lambda_{nb} r_{nb}^2}{2(1 - \rho_{nb})} + \frac{\lambda_{sb} r_{sb}^2}{2(1 - \rho_{sb})}$$

or, substituting λ's and ρ's,

$$D = 0.1585 r_{eb}^2 + 0.1996 r_{wb}^2 + 0.07115 r_{nb}^2 + 0.04 r_{sb}^2$$

It is stated in the problem that east and west effective red are equal and north and south effective reds are equal, so let

$$r_{ew} = r_{eb}$$
$$= r_{wb}$$
$$r_{ns} = r_{nb}$$
$$= r_{sb}$$

and by definition (with a 60-sec cycle length and zero lost times)

$$r_{ns} = 60 - r_{ew}$$

Substituting this into the total delay expression gives

$$D = 0.1585 r_{ew}^2 + 0.1996 r_{ew}^2 + 0.07115(60 - r_{ew})^2 + 0.04(60 - r_{ew})^2$$

or

$$D = 0.46925 r_{ew}^2 - 13.338 r_{ew} + 400.14$$

At the minimum of total delay $dD/dr_{ew} = 0$. Thus differentiating yields

$$\frac{dD}{dr_{ew}} = 0.9385 r_{ew} - 13.338$$

$$= 0$$

Therefore,

$$r_{ew} = \underline{\underline{14.2 \text{ sec}}} \quad \text{and} \quad r_{ns} = 60 - 14.2 = \underline{\underline{45.8 \text{ sec}}}$$

and

$$D = 0.1585(14.2)^2 + 0.1996(14.2)^2 + 0.07115(45.8)^2 + 0.04(45.8)^2$$
$$= 305.6 \text{ veh-sec/cycle}$$

Therefore, optimal signal timing for this problem is to allocate 45.8 sec of effective green (14.2 sec of effective red) to the east–west approaches and 14.2 sec of effective green (45.8 sec of effective red) to the north–south approaches. By substituting the optimal effective reds into the total delay equation, the minimum total intersection delay per cycle is found to be 305.6 vehicle-seconds. Finally, it can be readily shown that for the 45.8 to 14.2 sec signal timing, the earlier assumption that capacity exceeds arrivals is satisfied for all approaches.

NOMENCLATURE
FOR
CHAPTER 5

c time for signal to complete one cycle

d average vehicle delay per cycle assuming deterministic arrivals

d' average vehicle delay per cycle assuming random arrivals

d_m maximum delay of any vehicle

D total vehicle delay per cycle

g effective green

h headway

k traffic density

k_j jam density

k_m traffic density at maximum flow

l roadway length

n number of vehicles

P_q proportion of the signal cycle with a queue

P_s proportion of stopped vehicles

q traffic flow

q_m maximum traffic flow

Q length of queue

$\overline{Q}$ average length of queue

Q_m maximum number of vehicles in the queue

r effective red

s saturation flow

t time

$\bar{t}$ average waiting time in the system

t_0 time after the start of effective green until queue dissipation

u space mean speed

u_f free flow speed

u_m speed at maximum flow

$\bar{u}_t$ time mean speed

$\overline{w}$ average waiting time in the queue

λ arrival rate

μ departure rate

ρ traffic intensity

REFERENCES

1. "Traffic Flow Theory: A Monograph", Transportation Research Board Special Report 165, Washington, D.C. 1975.

2. L. A. Pipes, "Car Following Models and the Fundamental Diagram of Road Traffic," *Transportation Research*, Vol. 1, No. 1, 1967.

3. D. R. Drew, "Deterministic Aspects of Freeway Operations and Control," *Highway Research Record*, 99, 1965.

4. F. A. Haight, Counting Distributions for Renewal Processes, *Biometrica*, Vol. 52, Nos. 3, 4, 1965.

5. E. Gelenbe and G. Pujolle, *Introduction to Queuing Networks*, John Wiley & Sons, New York, 1987.

6. F. A. Haight, *Mathematical Theories of Traffic Flow*, Academic Press, New York, 1963.

7. F. V. Webster, "Traffic Signal Settings," Road Research Technical Paper No. 39, Great Britain Road Research Laboratory, London, 1958.

8. R. E. Allsop, "Delay at a Fixed Time Traffic Signal, I: Theoretical Analysis," *Transportation Science*, Vol. 6, No. 3, 1972.

9. Transportation Research Board Special, *Highway Capacity Manual*, Report 209, Washington, D.C., 1985.

10. R. A. Vincent, A. I. Mitchell, and D. I. Robertson, "User Guide to TRANSYT Version 8," Transport and Road Research Laboratory Report 888, Crowthorne, England, 1980.

11. F. L. Mannering, "Compensating Variation Approach to Optimal Signal Timings," presented at the spring meeting of the Operations Research Society of America, Washington, D.C., 1988.

PROBLEMS

5.1. On a lightly congested roadway, a transportation student finds that 60 percent of the headways between cars are 13 sec or greater. If the student decides to count traffic in 30-sec intervals, estimate the probability of the student's counting exactly four vehicles in an interval.

5.2. On a lightly traveled highway, a transportation student counts cars over 120 in 20-sec intervals. It is noted that no cars arrive in 18 of the intervals. Approximate the number of intervals in which exactly three cars arrive.

5.3. For the data collected in Problem 5.2, estimate the percentage of headways that (1) will be 10 sec or greater, and (2) will be less than 6 sec.

5.4. A toll booth on a turnpike is open from 8:00 a.m. to 12 midnight. Vehicles start arriving at 7:45 a.m. at a uniform deterministic rate of 6 per minute until 8:15 a.m. and from then on at 2 per minute. If the vehicles are serviced at a uniform deterministic rate of 6 per minute, determine:

(a) When the morning queue will dissipate.

(b) Total delay.

(c) The longest queue length.

(d) The longest customer delay under FIFO.

(e) The longest customer delay under LIFO.

5.5. At a parking lot, vehicles arrive according to a Poisson process and are servied at a uniform deterministic rate at a single station. The mean arrival rate is 4 veh/min and the mean service rate is 5 veh/min.

(a) Determine the average length of queue, time spent in the system, and waiting time spent in queue.

(b) If the service times become exponentially distributed, find the revised values of the three factos in part (a).

5.6. Vehicles arrive (Poisson arrivals) at a toll booth with a mean arrival rate of 2 veh/min. The toll booth operator processes vehicles at a uniform deterministic rate of 1 vehicle every 20 sec. What is the average length of queue, time spent in the system, and waiting time spent in the queue?

5.7. A transportation student buys a gasoline service station in a residential area. To increase the station's exposure, the student decides to sell gasoline at 1960s prices (25 cents per gallon) for 1 hour (9:00 to 10:00 a.m.). Vehicles start arriving at 8:45 a.m. at a uniform deterministic rate of four per minute and continue to arrive at this rate until 9:15 a.m. From 9:15 to 10:00 a.m. the arrival rate becomes 8 per minute.

(a) If the student opens late and services the cars from 9:15 to 10:00 a.m. at a constant rate of 11 cars per minute, determine total delay, maximum queue length, and longest customer delay assuming FIFO and LIFO.

(b) If an irate neighbor told the student to eliminate all queues by 9:45 a.m., what service rate would the student have to maintain (assume all other conditions are the same as before)?

5.8. A ferryboat queuing lane holds 30 vehicles. If the service rate is a uniform deterministic 4 vehicles per minute and service begins when the lane reaches capacity, what is the uniform deterministic arrival rate if the vehicle queue dissipates 30 min after vehicles begin to arrive?

5.9. At a toll booth, vehicles arrive and are serviced at uniform deterministic rates, λ and μ, repsectively. The arrival rate is 2 veh/min. Service begins 13 min after the arrival of the first vehicle and the queue dissipates t min after the arrival of the first vehicle. If the number of vehicles that must actually wait in a queue is x, develop an expression for determining service rates in terms of x.

5.10. Vehicles arrive at an amusement park at a uniform deterministic of 4 veh/min. If uniform deterministic service begins 30 min after the first arrival and the total delay is 3600 veh-min, how long after the first arrival will it take the queue to dissipate?

5.11. An intersection has a saturation flow of 1500 veh/hr and cars arrive at the approach at the rate of 800 veh/hr. It is controlled by a pretimed signal with a cycle length of 60 sec and a $D/D/1$ queuing regime holds. Local standards dictate that signals should be set such that all of the queues dissipate 10 sec before the end of the effective green portion of the cycle.

Assuming that approach capacity exceeds arrivals, determine the maximum length of effective red that will satisfy local standards.

5.12. An approach at a pretimed signal has 30 sec of effective red and a $D/D/1$ regime holds. The total delay is 83.33 veh-sec/cycle and the saturation flow is 1000 veh/hr. If the capacity of the approach equals the number of arrivals per cycle, determine the approach flow and cycle length.

5.13. An approach to a pretimed signal has 25 sec of effective green time in a 60-sec cycle. The approach volume is 500 veh/hr and the saturation flow is 1400 veh/hr. Calculate the average vehicle delay using the $D/D/1$ regime, Webster's formula, and Allsop's formula.

5.14. A transportation student observes an approach to a pretimed signalized intersection and notes that the maximum number of vehicles in a queue is 8 for a given cycle. If the saturation flow is 1440 veh/hr and the effective red time is 40 sec, how much time will it take this queue to dissipate after the start of the effective green (assuming that approach capacity exceeds arrivals and a $D/D/1$ queuing regime.)

5.15. Recent computatons at an approach of a pretimed signalized intersection indicate that the average delay per vehicle is 16.6 sec. (as calculated by Allsop's expression). The volume to capacity ratio is 0.8, the saturation flow is 1600 veh/hr, and the effective green time is 50 sec. If the uniform delay (assuming a $D/D/1$ regime) is 11.25 sec/veh, determine the arrival flow (in veh/hr) and the cycle length.

5.16. An approach to a pretimed signal has 25 sec of effective green, a saturation flow of 1300 veh/hr, and a volume to capacity ratio less than one. If the cycle length is 60 sec and Allsop's delay formula estimates a delay that is 34 sec/veh higher than that estimated by using the $D/D/1$ delay formula, determine the vehicle arrival rate.

5.17. An approach to a pretimed signal with a 60-sec cycle has 8.9 vehicles in the queue at the beginnning of the effective green. Four of the 8.9 vehicles in the queue are left over from the previous cycle. The saturation flow of the approach is 1500 veh/hr, total delay for the cycle is 5.78 veh-min, and at the end of the effective green 2 vehicles are in the queue. Determine the arrival rate assuming that it is unchanged over the duration of observation period (i.e., from the beginning to the end of the 5.78 veh-min delay cycle). Use a $D/D/1$ queuing regime.

5.18. At the beginning of an effective red, vehicles are arriving at an approach at the rate of 500 veh/hr and 16 vehicles are left in the queue from the previous cycle. However, due to the end of a major sporting event, the arrival rate is continuously increasing at a constant rate of 200 veh/hr/min (i.e., after 1 min the arrival rate will be 700 veh/hr; after 2 min, 900 veh/hr, etc.). The saturation flow of the approach is 1800 veh/hr, the cycle length is 60 sec, and the effective green is 40 sec. Determine the total vehicle delay until complete queue dissipation (assume a $D/D/1$ queuing regime).

5.19. The saturation flow for an intersection approach is 3600 veh/hr. At the beginning of a cycle (effective red) no vehicles are queued. The signal is timed so that when the queue is 13 vehicles long, the effective green begins. If the queue dissipates 8 sec before the end of the cycle and the cycle length is 60 sec, what is the arrival rate (assume a $D/D/1$ queuing regime)?

Chapter Six

Applied Traffic Analysis:
Basic Elements

6.1 INTRODUCTION

Although the fundamental elements of traffic analysis presented in Chapter 5 are theoretically valid, the actual application of these fundamentals requires a more elaborate definition of terms. Two examples of the practice-oriented need for more elaborate definitions come to mind; capacity and flow (i.e., in *vehicles* per hour). In Chapter 5, capacity (q_m) is defined simply as the highest rate of flow that the highway is capable of supporting. For applied traffic analysis, a consistent and reasonably precise method of determining capacity must be developed within this definition. Since the capacity of a roadway section is a function of factors such as roadway type (e.g., freeway, arterial, or rural road), design speed, the number of lanes and widths of lanes and shoulders, the method of capacity determination clearly must account for a wide variety of physical roadway features.

The next concern is traffic flow, which is defined in Chapter 5 in typical units of vehicles per hour. Two practical issues arise in this regard. First, in many cases, actual traffic streams consist of a variety of vehicle types with substantially different performance characteristics that are likely to be magnified by changing roadway geometrics (e.g., grades, as discussed in Chapter 2). Thus traffic must not only be defined in terms of vehicles per unit time but also in terms of vehicle composition, since it is clear that a 1500-veh/hr traffic flow, consisting of 100 percent automobiles, will differ significantly in terms of operating speed and density when compared with a 1500-veh/hr traffic flow that consists of 50 percent automobiles and 50 percent heavy trucks. The other flow-related issue relates to the temporal distribution of traffic. Traditionally, the analysis of traffic on a roadway focuses on the most critical condition, which is the most congested hour within a 24-hour daily period (the temporal distribution of traffic will be discussed in more detail in Section 6.7). However, within this most congested or *peak hour*, traffic flow is likely to be nonuniform, as illustrated in Fig. 6.1. Since in most instances the focus of traffic analysis is actually on the most congested time period within a peak hour, some method of defining and measuring the nonuniformity of flow is obviously needed.

The objective of applied traffic analysis is to provide a practical method of quantitatively measuring the degree of traffic congestion and being able to relate this to the degree of safety and efficiency provided by the highway. The following sections of this chapter discuss and demonstrate accepted standards for applied traffic analysis.

6.2 LEVEL OF SERVICE

A qualitative measure describing traffic operational conditions and their perception by drivers is needed to assess the degree of congestion on a highway facility. Such a measure is referred to as a *level of service* and is intended to capture factors such as speed and travel time, freedom to maneuver, and safety. Current practice designates level of service ranging from A to F, with level-of-service A representing the best operating conditions and level-of-service F the worst. The *Highway Capacity Manual* [Transportation Research Board 1985] defines level of service for freeways (LOS) as follows:

Level-of-Service A   LOS A represents free flow. Individual users are virtually unaffected by the presence of others in the traffic stream. Freedom to select desired speeds and to maneuver within the traffic stream is extremely high. The general level of comfort and convenience provided to the motorist or passenger is excellent.

Level-of-Service B   LOS B is in the range of stable flow, but the presence of other users in the traffic stream begins to be noticeable. Freedom to select desired speeds is relatively unaffected, but there is a slight decline in the freedom to maneuver within the traffic stream from LOS A. The level of

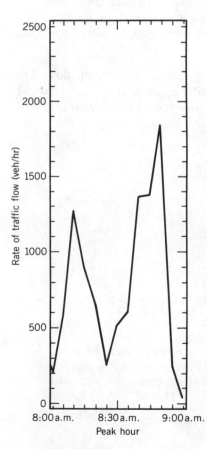

FIGURE 6.1
Example of nonuniform flow over a peak hour.

comfort and convenience provided is somewhat less than at LOS A, because the presence of others in the traffic stream begins to affect individual behavior.

Level-of-Service C   LOS C is in the range of stable flow but marks the beginning of the range of flow in which the operation of individual users becomes significantly affected by the interactions with others in the traffic stream. The selection of speed is now affected by the presence of others, and maneuvering within the traffic stream requires substantial vigilance on the part of the user. The general level of comfort and convenience declines noticeably at this level.

Level-of-Service D   LOS D represents high density, but stable, flow. Speed and freedom to maneuver are severely restricted, and the driver experiences a generally poor level of comfort and convenience. Small increases in traffic flow will generally cause operational problems at this level.

Level-of-Service E   LOS E represents operating conditions at or near the capacity level. All speeds are reduced to a low, but relatively uniform, value. Freedom to maneuver within the traffic stream is extremely difficult, and it is generally accomplished by forcing a vehicle to "give way" to accommodate such maneuvers. Comfort and convenience levels are extremely poor, and driver frustration is generally high. Operations at this level are usually unstable, because small increases in flow or minor perturbations within the traffic stream will cause breakdowns in flow.

Level-of-Service F   LOS F is used to define forced or breakdown flow. This condition exists wherever the amount of traffic approaching a point exceeds the amount that can traverse the point. Queues form behind such locations. Operations within the queue are characterized by stop-and-go traffic *waves*,

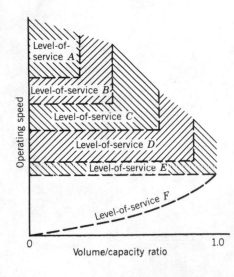

FIGURE 6.2
Relationship between level of service, speed, and volume-to-capacity ratio.
Source: Highway Research Board, High-way Capacity Manual, Special Report 87; National Research Council, Washington, D.C., 1965.

Level-of-Service A

Level-of-Service D

Level-of-Service B

Level-of-Service E

Level-of-Service C

Level-of-Service F

FIGURE 6.3
Illustration of freeway level-of-service (A to F). (Reproduced by permission from Transportation Research Board, *Highway Capacity Manual*, Special Report 209, National Research Council, Washington, D.C., 1985).

and they are extremely unstable. Vehicles may progress at reasonable speeds for several hundred feet or more, then be required to stop in a cyclic fashion. LOS F is used to describe the operating conditions within the queue, as well as the point of traffic flow breakdown. It should be noted, however, that in many cases operating conditions of vehicles discharged from the queue may be quite good. Nevertheless, it is the point at which arrival flow exceeds discharge (or service flow) that causes the queue to form; LOS F is an appropriate designation for such points.

These level-of-service designations can be illustrated conceptually within the context of a speed-flow diagram, as presented in Chapter 5 (Fig. 5.3). Figure 6.2 provides such a diagram with the volume-to-capacity ratio (v/c or equivalently q/q_m using the notation of Chapter 5) replacing flow on the x-axis [Highway Research Board 1965]. This figure, and the preceding level-of-service definitions, provide an important traffic analysis fundamental in that roadway capacity (which will be shown to be a function of prevailing traffic and physical roadway features), will always be reached when the facility is operating at level-of-service E. Clearly, a roadway operating at capacity does not provide a desirable level of service from the driver's perspective and is a situation that should be avoided in the design of a highway. Finally, to gain a visual perspective of level of service, Fig. 6.3 presents an illustration of traffic conditions on a freeway segment varying from level-of-service A through F.

6.3 BASIC DEFINITIONS

In determining the level of service of a highway segment, a few key definitions and associated notations must be well understood.

Hourly Volume Hourly volume is the actual hourly demand volume for the highway in vehicles per hour, given the symbol V. Generally, the highest 24-hour hourly volume (i.e., peak-hour volume) is used for V in traffic analysis computations.

Peak-Hour Factor The peak-hour factor accounts for the nonuniformity of traffic flow over the peak hour (as mentioned in Section 6.1). It is denoted PHF and is typically defined as the ratio of the hourly volume (V) to the maximum 15-min rate of flow (V_{15}) expanded to an hourly volume. Therefore,

$$\text{PHF} = \frac{V}{V_{15} \times 4} \tag{6.1}$$

Equation 6.1 indicates that the further the PHF is form unity, the more *peaked* or nonuniform the flow. For example, consider two roads both of which have a peak-hour volume, V, of 2000 veh/hr. However, the first road has 1000 vehicles arriving in the highest 15-min interval and the second road has 600 vehicles arriving in the highest 15-min interval. Clearly, the first road has a more nonuniform flow, and this is substantiated by the fact that its PHF of 0.5 [i.e., $2000/(1000 \times 4)$] is further from unity than the second road's PHF of 0.83 [i.e., $2000/(600 \times 4)$].

Service Flow   Service flow is the actual rate of flow for the peak 15-min period expanded to an hourly volume and expressed in vehicles per hour. Service flow is denoted SF and is defined as

$$SF = \frac{V}{PHF} \qquad (6.2)$$

or

$$SF = V_{15} \times 4 \qquad (6.3)$$

The above definitions apply to all basic highway types: freeways, multilane highways, and two-lane highways (one lane in each direction). However, there are a number of additional terms that must be introduced before highway level-of-service analysis can be undertaken. These are best defined within specific highway types as presented in the following section.

6.4 BASIC FREEWAY SEGMENTS

A basic freeway segment is defined as a section of a divided highway having two or more traffic lanes in one direction, full access control, and traffic that is unaffected by merging or diverging movements near ramps or lane additions or lane deletions. It is important to note that capacity analysis for divided highways focuses on the flow of traffic in one direction only. This is a logical approach, because the concern is to measure the true level of traffic congestion and, since it is unlikely that during a given hour both highway directions will reach their peak flows, consideration of traffic volumes in both directions is likely to seriously understate the overall level of traffic congestion.

To begin assessment of a roadway's level of service, *ideal* roadway conditions must be specified. Recall that in the introduction of this chapter the effect of factors such as different vehicle performance characteristics and roadway design features, on traffic flow, was discussed qualitatively. In practice, the effect of such factors on traffic flow is measured quantitatively, relative to some conditions that are considered ideal. For freeways, ideal conditions can be categorized as those relating to lane widths and/or lateral clearances, the effects of heavy vehicles, and driver population characteristics. Studies have shown that the ideal lane width is 12 ft and objects (e.g., telephone poles or barriers) should be no closer than 6 ft from the traveled pavement (at the roadside or median). Also, under ideal conditions, there should be passenger cars only in the traffic stream with no heavy vehicles such as buses or trucks, and the driver population should be weekday drivers or commuters who, due to their presumed familiarity with traffic and roadway conditions, will behave so as to maximize the efficient flow of traffic.

With the notion of ideal conditions, the term *maximum service flow*, MSF_i, can be defined for a given level of service i as the highest service flow that can be achieved, while maintaining the specified level-of-service i, assuming ideal road-way conditions. Since ideal conditions specify the presence of passenger cars only, and it is desirable to make the maximum service flow rate independent of the number of highway lanes, MSF_i is in units of passenger cars per hour per lane (pcphpl). Maximum flow rates and accepted level-of-service criteria for freeway segments with design speeds (as discussed in Chapter 3) of 70, 60, and 50 mph are given in Table 6.1 along with approximate densities and operating speeds for each level of service. Also, note that each level of service has a maximum volume-to-capacity ratio that corresponds to a maximum service flow rate. Within this context, one of the basic relationships underlying Table 6.1 can be expressed as

$$MSF_i = c_j \times \left(\frac{v}{c}\right)_i \qquad (6.4)$$

where MSF_i is the maximum service flow rate per lane for level-of-service i under ideal conditions in pcphpl, $(v/c)_i$ is the maximum volume-to-capacity ratio associated with level-of-service i (see Table 6.1), and c_j is the capacity under ideal conditions for a freeway with design speed j. Also, c_j is 2000 pcphpl for 60- and 70-mph design speeds and 1900 pcphpl for 50-mph design speeds. Note that the value of c_j equals the maximum service flow rate at LOS E in Table 6.1 as suggested by Fig. 6.2 [i.e., $(v/c)_E = 1$]. Also, in applying Eq. 6.4 with $(v/c)_i$ multiplied by c_j, note that the values of MSF presented in Table 6.1 are rounded to the nearest 50 pcphpl.

6.4.1 Service Flow Rates and Level of Service

The concept of a maximum service flow rate provides an important bench mark for determining a highway's level of service but, since ideal conditions are seldom realized in practice, some method of converting the maximum service flow rate into an equivalent service flow rate (which accounts for actual prevailing conditions) is needed. Once this is achieved, the highest service flow rate at prevailing conditions for a given level of service (SF_i) can be related to the service flow rate obtained from actual vehicle counts (i.e., SF in Eqs. 6.2 and 6.3) to determine the highway's level of service (as will soon be demonstrated by example). In calculating service flow rates under prevailing conditions, correction factors are used along with the number of directional lanes giving

$$SF_i = MSF_i \times N \times f_w \times f_{HV} \times f_p \qquad (6.5)$$

TABLE 6.1
Levels-of-Service Criteria for Freeways

LOS	Density (PC/mi/ln)	70-mph Design Speed			60-mph Design Speed			50-mph Design Speed		
		Speed[b] (mph)	v/c	MSF[a] (PCPHPL)	Speed[b] (mph)	v/c	MSF[a] (PCPHPL)	Speed[b] (mph)	v/c	MSF[a] (PCPHPL)
A	≤ 12	≥ 60	0.35	700	—	—	—	—	—	—
B	≤ 20	≥ 57	0.54	1,100	≥ 50	0.49	1,000	—	—	—
C	≤ 30	≥ 54	0.77	1,550	≥ 47	0.69	1,400	≥ 43	0.67	1,300
D	≤ 42	≥ 46	0.93	1,850	≥ 42	0.84	1,700	≥ 40	0.83	1,600
E	≤ 67	≥ 30	1.00	2,000	≥ 30	1.00	2,000	≥ 28	1.00	1,900
F	> 67	< 30	c	c	< 30	c	c	< 28	c	c

Source: Transportation Research Board, *Highway Capacity Manual*, Special Report 209, National Research Council, Washington, D.C., 1985.
[a]Maximum service flow rate per lane under ideal conditions.
[b]Average travel speed.
[c]Highly variable, unstable.
Note: All values of MSF rounded to the nearest 50 pcph.

where SF_i is the service flow rate for level of service i under prevailing conditions for N lanes (in one direction) in vehicles per hour, N is the number of directional lanes, f_w is a factor to adjust for the effects of less than ideal lane widths and/or lateral clearances, f_{HV} is a factor to adjust for the effect of nonpassenger cars in the traffic stream (i.e., heavy vehicles such as trucks, buses, and recreational vehicles) and f_p is a factor to adjust for the effect of nonideal driver populations (i.e., nonregular travelers). By combining Eqs. 6.4 and 6.5, another useful equation is obtained:

$$SF_i = c_j \times (v/c)_i \times N \times f_w \times f_{HV} \times f_p \tag{6.6}$$

Equations 6.4, 6.5, and 6.6 form the basis for freeway capacity/congestion analysis.

TABLE 6.2
Adjustment Factor for Restricted Lane Width and Lateral Clearance for Freeways

Distance from Traveled Pavement[a] (ft)	Adjustment Factor, f_w							
	Obstructions on One Side of the Roadway				Obstructions on Both Sides of the Roadway			
	Lane Width (ft)							
	12	11	10	9	12	11	10	9
	4-Lane Freeway (2 Lanes each direction)							
≥ 6	1.00	0.97	0.91	0.81	1.00	0.97	0.91	0.81
5	0.99	0.96	0.90	0.80	0.99	0.96	0.90	0.80
4	0.99	0.96	0.90	0.80	0.98	0.95	0.89	0.79
3	0.98	0.95	0.89	0.79	0.96	0.93	0.87	0.77
2	0.97	0.94	0.88	0.79	0.94	0.91	0.86	0.76
1	0.93	0.90	0.85	0.76	0.87	0.85	0.80	0.71
0	0.90	0.87	0.82	0.73	0.81	0.79	0.74	0.66
	6- or 8-Lane Freeway (3 or 4 Lanes each direction)							
≥ 6	1.00	0.96	0.89	0.78	1.00	0.96	0.89	0.78
5	0.99	0.95	0.88	0.77	0.99	0.95	0.88	0.77
4	0.99	0.95	0.88	0.77	0.98	0.94	0.87	0.77
3	0.98	0.94	0.87	0.76	0.97	0.93	0.86	0.76
2	0.97	0.93	0.87	0.76	0.96	0.92	0.85	0.75
1	0.95	0.92	0.86	0.75	0.93	0.89	0.83	0.72
0	0.94	0.91	0.85	0.74	0.91	0.87	0.81	0.70

Source: Transportation Research Board, *Highway Capacity Manual*, Special Report 209, National Research Council, Washington, D.C., 1985.
[a] Certain types of obstructions, high-type median barriers in particular, do not cause any deleterious effect on traffic flow. Judgment should be exercised in applying these factors.

6.4.2 Lane Width and/or Lateral Clearance Adjustment

When lane widths are narrower than the ideal 12 ft and/or obstructions (e.g., barriers or telephone poles) are closer than 6 ft from the edge of the traveled pavement (at the roadside or at the median), the adjustment factor f_w is used to reflect prevailing conditions. Such an adjustment is needed, since narrow lanes and obstructions close to traveled pavement cause traffic to slow as a result of reduced psychological comfort and limits on driver maneuvering and accident avoidance options. This in turn leads to an effective reduction in roadway capacity relative to the capacity that would be available if ideal roadway conditions were provided.

The adjustment factors used in current practice are presented in Table 6.2. Although the definition of lane width is unambiguous, some elaboration of what is meant by an obstruction is warranted. An obstruction is a roadside or median object that can be either continuous (e.g., a retaining wall or barrier) or periodic (e.g., a light post or telephone poles). Table 6.2 provides corrections for obstructions on one side of the roadway (either median or roadside objects) and for obstructions on both sides (both median and roadside). For the case where obstructions are on both sides of the roadway and the distances from the traveled pavement to the objects are unequal (e.g., 2 ft to roadside obstructions and 4 ft to median obstructions), the average distance is used to arrive at the f_w value from Table 6.2.

As an example of using Table 6.2, consider a 6-lane freeway with 11-ft lanes and obstructions from the traveled pavement of 4 ft for the roadside edge and 0 ft (e.g., the barrier immediately adjacent) from the median edge. The f_w factor is found to be 0.92 from Table 6.2 indicating that 8 percent of the ideal roadway capacity is lost due to nonideal lane widths and lateral clearances.

6.4.3 Heavy Vehicle Adjustment

Trucks, buses, and recreational vehicles have performance characteristics (slow acceleration or inferior braking) and dimensions (length, height, and width) that have an adverse effect on a roadway's capacity. Recall that ideal conditions assume that no heavy vehicles are present in the traffic stream and, when prevailing conditions indicate the presence of heavy vehicles, the adjustment factor f_{HV} is used to translate ideal to prevailing conditions. The f_{HV} correction term is found using a two-step process. The first step is to determine the passenger car equivalent (pce) for each truck, bus, and/or recreational vehicle in the traffic stream. These values represent the number of passenger cars that would consume the same amount of roadway capacity as a single truck, bus, or recreational vehicle. These passenger car equivalents are denoted E_T for truck, E_B for buses and E_R for recreational vehicles, and are a function of the type of terrain (e.g., roadway grades), since steep grades will tend to magnify the performance inferiority of heavy vehicles as well as the sight-distance problems caused by their larger dimensions (i.e., visibility of drivers in vehicles following

TABLE 6.3
Passenger Car Equivalents for Freeways

Factor	Type of Terrain		
	Level	Rolling	Mountainous
E_T for trucks	1.7	4.0	8.0
E_B for buses	1.5	3.0	5.0
E_R for recreational vehicles	1.6	3.0	4.0

Source: Transportation Research Board, *Highway Capacity Manual*, Special Report 209, National Research Council, Washington, D.C., 1985.

heavy vehicles). For roadway segments where no single grade of 3 percent or greater is longer than $\frac{1}{2}$ mile or longer than 1 mile for grades less than 3 percent, passenger car equivalency factors can be obtained from Table 6.3 with terrain types defined as [Transportation Research Board 1985].

Level Terrain   Any combination of grades and horizontal or vertical alignment permitting heavy vehicles to maintain approximately the same speed as passenger cars; this generally includes short grades of no more than 1 to 2 percent.

Rolling Terrain   Any combination of grades and horizontal or vertical alignment causing heavy vehicles to reduce their speeds substantially below those of passenger cars but not causing them to operate at their maximum speed for given roadway geometrics (i.e., $F_{net}(V) = 0$ due to steep grades as illustrated in Fig. 2.6) for any significant length of time.

Mountainous Terrain   Any combination of grades and horizontal or vertical alignment causing heavy vehicles to operate at their maximum speed for given roadway geometrics for significant distances or at frequent intervals.

If the roadway has grades greater than 3 percent that are longer than $\frac{1}{2}$ mile or grades less than 3 percent that are longer than 1 mile, the values in Table 6.3 are no longer valid. In such cases, equivalency factors can be obtained from more detailed tables, as presented in the *Highway Capacity Manual* [Transportation Research Board 1985].

Once the appropriate equivalency factors have been obtained, the following equation is applied to arrive at the heavy vehicle correction factor f_{HV}:

$$f_{HV} = \frac{1}{1 + P_T(E_T - 1) + P_B(E_B - 1) + P_R(E_R - 1)} \qquad (6.7)$$

where the P's are the proportions of heavy vehicles in the traffic stream and the E's are the equivalency factors as read from Table 6.3.

As an example of how a heavy vehicle correction factor is computed, consider a freeway on rolling terrain with 10 percent of the traffic flow being trucks, 7 percent buses, and 3 percent recreational vehicles. The corresponding equivalency factors for this roadway are $E_T = 4.0$, $E_B = 3.0$, and $E_R = 3.0$, as indicated in Table 6.3. Also, from the given percentages of heavy vehicles in the traffic stream, $P_T = 0.1$, $P_B = 0.07$, and $P_R = 0.03$. Substituting these values into Eq. 6.7 gives $f_{HV} = 0.67$ or a 33 percent reduction in roadway capacity relative to the ideal condition of having no heavy vehicles in the traffic stream.

6.4.4 Driver Population Adjustment

Under ideal conditions, the traffic stream is assumed to consist of regular weekday drivers and commuters. Such drivers have a high familiarity with the freeway system and generally maneuver and respond to the maneuvers of other drivers safely and predictably. However, in many situations, the traffic stream has a driver population that is less familiar with local highway conditions (e.g., weekend drivers or drivers in recreational areas) so that a substantial loss of roadway capacity, relative to the ideal driver population, can result.

To account for the composition of the driver population, the f_p adjustment factor is used and its recommended values are given in Table 6.4. Note that for nonideal driver populations (i.e., "other" in Table 6.4), the loss in roadway capacity can vary from 10 to 25 percent. This nonideal correction is dependent on local conditions such as highway conditions and surrounding environment (i.e., possible driver distractions such as scenic views, and so on). When nonideal driver populations are present, engineering judgment and local data should be used to determine the appropriate driver population correction term (for further information, see [Transportation Research Board 1985].)

TABLE 6.4
Adjustment Factor for Driver Population on Freeways

Traffic Stream Type	Factors, f_p
Weekdays or commuter	1.0
Other	0.75–0.90[a]

Source: Transportation Research Board, *Highway Capacity Manual*, Special Report 209, National Research Council, Washington, D.C., 1985.
[a] Engineering judgment and/or local data must be used in selecting an exact value.

6.4.5 Freeway Traffic Analysis

With all terms in Eqs. 6.4, 6.5, and 6.6 defined, these equations can now be applied to determine freeway level of service and freeway capacity. The manner in which this determination is undertaken is best demonstrated by example.

Example 6.1

A six-lane urban freeway (three lanes in each direction) is on rolling terrain with a 70-mph design speed, 10-ft lanes, with obstructions 2 ft from both roadside and median edges. A directional weekday peak-hour volume of 2200 vehicles is observed with 700 vehicles arriving in the most congested 15-min period. If the traffic stream has 12 percent trucks, 10 percent buses, and 2 percent recreational vehicles, determine the level of service.

Solution

The approach that will be taken to determine level of service will be to compute the volume-to-capacity ratio (v/c) of the roadway and to compare it with the maximum volume to capacity ratios for specified levels of service, as given in Table 6.1. To arrive at the roadway's volume-to-capacity ratio, Eq. 6.6 is rearranged giving

$$v/c = \frac{SF}{c_j \times N \times f_w \times f_{HV} \times f_p}$$

where, from Eq. 6.3,

$$SF = V_{15} \times 4$$

$$= 700 \times 4$$

$$= 2800 \text{ veh/hr}$$

and

$c_j = 2000$ pcphpl (70-mph design speed)
$N = 3$ (given)
$f_p = 1.0$ (weekday, Table 6.4)
$f_w = 0.85$ (10-ft lanes, obstructions 2 ft on both sides, assumed with 2-ft shoulder, Table 6.2)
$E_T = 4.0$, $E_B = 3.0$, $E_R = 3.0$ (rolling terrain, Table 6.3)

From Eq. 6.7, we obtain

$$f_{HV} = \frac{1}{1 + 0.12(4 - 1) + 0.10(3 - 1) + 0.02(3 - 1)}$$

$$= 0.625$$

So, substituting, we find that

$$v/c = \frac{2800}{2000 \times 3 \times 0.85 \times 0.625 \times 1.0}$$

$$= 0.878$$

which gives a LOS D from Table 6.1, since the maximum v/c for LOS C is 0.77 and the maximum v/c for LOS D is 0.93 (i.e., 0.77 < 0.878 < 0.93).

Example 6.2

Considering the conditions described in Example 6.1, how many additional vehicles can be added to the peak hour before the freeway reaches capacity?

Solution

To solve this problem, the highest service flow that the roadway can sustain must be computed. Since roadway capacity occurs at LOS E, and the highest volume-to-capacity ratio under LOS E is 1.0 (Table 6.1, with volume equal to capacity) the highest service flow can be calculated from Eq. 6.6,

$$SF_E = v/c_E \times c_j \times N \times f_w \times f_{HV} \times f_p$$

Using $v/c_E = 1.0$, and all other terms as in Example 6.1, we obtain

$$SF_E = 1.0 \times 2000 \times 3 \times 0.85 \times 0.625 \times 1.0$$

$$= 3187.5 \text{ veh/hr}$$

Recall that service flow is based on the highest 15-min volume in the peak hour. To determine the number of vehicles that can be added to the entire peak hour, service flow must be converted to an equivalent hourly volume. By rearranging Eq. 6.2, we get

$$V = SF \times PHF$$

where

$$PHF = \frac{V}{V_{15} \times 4}$$

or, since initial $V = 2200$ veh/hr (given) and $V_{15} = 700$ vehicle (given),

$$PHF = \frac{2200}{700 \times 4}$$
$$= 0.786$$

so that

$$V = 3187.5 \times 0.786$$
$$= 2505 \text{ veh/hr}$$

Assuming that all heavy vehicle percentages and traffic volume peaking remains the same, this means that 305 additional vehicles ($2505 - 2200$) can be added to the peak hour before the freeway reaches capacity.

6.5 MULTILANE HIGHWAYS

Multilane highways are those highways that are below freeway standards because they do not have full control of access and, in some instances, opposing directional lane groups are not divided by a median. In this book, attention will be directed toward multilane highways in rural and suburban settings only. The reader is referred to the *Highway Capacity Manual* [Transportation Research Board 1985] for the analysis of multilane highways (i.e., arterials) in urbanized areas.

Level-of-service analysis for multilane highways closely parallels that for freeways with ideal conditions, lane capacities by design speed (i.e., $c_{70} = c_{60} = 2000$ pcphpl, $c_{50} = 1900$ pcphpl), and most of the basic equations being the same. A notable exception is that Eq. 6.5 is now replaced by

$$SF_i = MSF_i \times N \times f_w \times f_{HV} \times f_p \times f_E \qquad (6.8)$$

The application of this equation is accomplished through the use of Tables 6.5, 6.6, 6.7, 6.8, and 6.9, which are used in essentially the same manner in which Tables 6.1, 6.2, 6.3, and 6.4, respectively, are used for freeway analysis. In this regard, note that Eq. 6.8 is identical to the freeway's Eq. 6.5 except for the additional adjustment factor for development environment and the type of

TABLE 6.5
Level-of-Service Criteria for Multilane Highways

Level of Service	Density (PC/mi/ln)	70-mph Design Speed			60-mph Design Speed			50-mph Design Speed		
		Speed[a] (mph)	v/c	MSF[b] (PCPHPL)	Speed[a] (mph)	v/c	MSF[b] (PCPHPL)	Speed[a] (mph)	v/c	MSF[b] (PCPHPL)
A	≤ 12	≥ 57	0.36	700	≥ 50	0.33	650	—	—	—
B	≤ 20	≥ 53	0.54	1,100	≥ 48	0.50	1,000	≥ 42	0.45	850
C	≤ 30	≥ 50	0.71	1,400	≥ 44	0.65	1,300	≥ 39	0.60	1,150
D	≤ 42	≥ 40	0.87	1,750	≥ 40	0.80	1,600	≥ 35	0.76	1,450
E	≤ 67	≥ 30	1.00	2,000	≥ 30	1.00	2,000	≥ 28	1.00	1,900
F	> 67	< 30	c	c	< 30	c	c	< 28	c	c

Source: Transportation Research Board, Highway Capacity Manual, Special Report 209, National Research Council, Washington, D.C., 1985.
[a] Average travel speed.
[b] Maximum rate of flow per lane under ideal conditions, rounded to the nearest 50 pcphpl.
[c] Highly variable.

TABLE 6.6
Adjustment Factor for Restricted Lane Width and Lateral Clearance for Multilane Highways

Distance from Edge of Traveled Way to Obstruction[a] (ft)	Adjustment Factor, f_w							
	Obstruction on One Side of Roadway[b]				Obstruction on Both Sides of Roadway[c]			
	Lane Width (ft)							
	12	11	10	9	12	11	10	9
4-Lane Divided Multilane Highways (2 lanes each direction)								
≥ 6	1.00	0.97	0.91	0.81	1.00	0.97	0.91	0.81
4	0.99	0.96	0.90	0.80	0.98	0.95	0.89	0.79
2	0.97	0.94	0.88	0.79	0.94	0.91	0.86	0.76
0	0.90	0.87	0.82	0.73	0.81	0.79	0.74	0.66
6-Lane Divided Multilane Highways (3 lanes each direction)								
≥ 6	1.00	0.96	0.89	0.78	1.00	0.96	0.89	0.78
4	0.99	0.95	0.88	0.77	0.98	0.94	0.87	0.77
2	0.97	0.93	0.87	0.76	0.96	0.92	0.85	0.75
0	0.94	0.91	0.85	0.74	0.91	0.87	0.81	0.70
4-Lane Undivided Multilane Highways (2 lanes each direction)								
≥ 6	1.00	0.95	0.89	0.77	NA	NA	NA	NA
4	0.98	0.94	0.88	0.76	NA	NA	NA	NA
2	0.95	0.92	0.86	0.75	0.94	0.91	0.86	NA
0	0.88	0.85	0.80	0.70	0.81	0.79	0.74	0.66
6-Lane Undivided Multilane Highways (3 lanes each direction)								
≥ 6	1.00	0.95	0.89	0.77	NA	NA	NA	NA
4	0.99	0.94	0.88	0.76	NA	NA	NA	NA
2	0.97	0.93	0.86	0.75	0.96	0.92	0.85	NA
0	0.94	0.90	0.83	0.72	0.91	0.87	0.81	0.70

Source: Transportation Research Board, *Highway Capacity Manual*, Special Report 209, National Research Washington, D.C., 1985.
[a]Use the average distance to obstruction on "both sides" where the distance to obstructions on the left and right differs.
[b]Factors for one-sided obstructions allow for the effect of opposing flow.
[c]Two-sided obstructions include one roadside and one median obstruction. Median obstruction may exist in the median of a divided multilane highway or in the center of an undivided highway which periodically divides to go around bridge abutments or other center objects.
NA = Not applicable; use factor for one-sided obstruction.

TABLE 6.7
Passenger Car Equivalents for Multilane Highways

Factor	Type of Terrain		
	Level	Rolling	Mountainous
E_T for trucks	1.7	4.0	8.0
E_B for buses	1.5	3.0	5.0
E_R for recreational vehicles	1.6	3.0	4.0

Source: Transportation Research Board, *Highway Capacity Manual*, Special Report 209, National Research Council, Washington, D.C., 1985.

TABLE 6.8
Adjustment Factor for the Type of Multilane Highway and Development Environment, f_E

Type	Divided	Undivided
Rural	1.00	0.95
Suburban	0.90	0.80

Source: Transportation Research Board, *Highway Capacity Manual*, Special Report 209, National Research Council, Washington, D.C., 1985.

multilane highway, f_E. The intent of this additional adjustment factor is twofold. First, it accounts for the superior traffic flow efficiency provided by divided (e.g., physical barrier separating opposing traffic) multilane highways relative to undivided (e.g., opposing flows separated only by centerline markings) multilane highways, which suffer reduced traffic flow efficiency due to the effects of interference from opposing traffic. Second, the f_E adjustment factor accounts for the fact that, since access is not controlled, suburban environments are likely to have more traffic stream interruption (turning movements, cross traffic, and so on) than their rural counterparts. Appropriate values of f_E are selected from Table 6.8.

It is also important to note that the lane width and lateral clearance factor, f_w, for multilane highways is dependent upon whether the highway is divided or undivided (see Table 6.6). This additional affect on f_w reflects the findings of recent research studies [Transportation Research Board 1985].

TABLE 6.9
Adjustment Factor for Driver Population on
Multilane Highways

Driver Population	Factor, f_p
Commuter or other regular users	1.00
Recreational or other nonregular users	0.75–0.90

Source: Transportation Research Board, *Highway Capacity Manual*, Special Report 209, National Research Council, Washington, D.C., 1985.

Example 6.3

A suburban multilane highway, four lanes (two lanes each direction), is undivided with telephone poles 8 ft from the pavement edge, 11-ft lanes, 60-mph design speed and a driving population of commuters. The highway is on rolling terrain and has peak-hour directional volume of 1600 vehicles (10 percent trucks, 5 percent buses, and 1 percent recreational vehicles) and a peak-hour factor of 0.90. Determine the level of service.

Solution

The level of service can be estimated by computing a volume-to-capacity ratio, as was done in Example 6.1. To begin, Eqs. 6.4 and 6.8 are combined to give

$$v/c = \frac{SF}{c_j \times N \times f_w \times f_{HV} \times f_E \times f_p}$$

where SF = V/PHF = 1600/0.9 = 1777.78 vph
c_j = 2000 pcphpl (60-mph design speed)
N = 2 (given)
f_w = 0.95 (11-ft lanes, no obstructions closer than 6 ft, Table 6.6)
f_E = 0.8 (suburban, undivided, Table 6.8)
f_p = 1.0 (commuters, Table 6.9)
E_T = 4.0, E_B = 3.0, E_R = 3.0 (rolling terrain, Table 6.7)

From Eq. 6.7, we obtain

$$f_{HV} = \frac{1}{1 + 0.10(4-1) + 0.05(3-1) + 0.01(3-1)}$$

$$= 0.704$$

Substituting to solve for v/c yields

$$v/c = \frac{1777.78}{2000 \times 2 \times 0.95 \times 0.704 \times 0.80 \times 1.0}$$

$$= 0.831$$

which gives LOS E (Table 6.5) for 60-mph design speed (i.e., $0.8 < 0.83 < 1.0$).

Example 6.4

The highway described in Example 6.3 is improved to a six-lane divided multi-lane facility (three lanes each direction). However, because of right-of-way restrictions, the lanes are only 9 ft wide, obstructions are on both sides 2 ft from the pavement edge, and the design speed had to be reduced to 50 mph. If all traffic-related conditions remain the same as those given in Example 6.3, determine the new level of service.

Solution

Again, the level of service is determined by computing the volume-to-capacity ratio as in Example 6.3. The terms used in the v/c computation are

where $f_{HV} = 0.704$ (as in Example 6.3)
$f_w = 0.75$ (9-ft lanes, 2-ft obstructions both sides, Table 6.6)
$f_E = 0.80$ (as in Example 6.3)
$f_p = 1.0$ (as in Example 6.3)
$c_j = 1900$ pcphpl (50-mph design speed)
$N = 3$ (given)

Solving for v/c gives

$$v/c = \frac{1777.78}{1900 \times 3 \times 0.75 \times 0.704 \times 0.8 \times 1.0}$$
$$= 0.738$$

which is, from Table 6.5, LOS D (i.e., $0.60 < 0.738 < 0.76$).

6.6 RURAL TWO-LANE HIGHWAYS

Two-lane highways are defined as two-lane roadways with one lane available to traffic in each direction. In terms of capacity determination, a key distinction between two-lane highways and the freeways and multilane highways previously discussed, is that slow-moving vehicles necessitate the use of an opposing traffic lane for passing operations. As presented in Chapter 3, the provision of sufficient sight distance to allow the necessary passing on two-lane roads is limited by roadway geometrics. It follows that any geometric features that restrict passing-sight distance will also restrict capacity. Moreover, because slower-moving vehicles cannot always be readily passed (i.e., passing can only occur in permitted zones and then only when no opposing traffic is present), the type of terrain not only impacts roadway geometrics but also affects the operating speeds of vehicles (see Chapter 2), which again impacts roadway capacity. Finally, since opposing traffic directly affects passing opportunities, the capacity of two-lane highways is viewed as a capacity in both directions as opposed to the one-direction approach used for the analysis of freeways and multilane highways. With traffic considered in both directions, the proportion of traffic flowing in each direction becomes a major consideration, as will be shown later.

With the above points in mind, ideal conditions can be defined for two-lane highways. In this book, attention will be given only to rural two-lane highways, and the reader is referred to the *Highway Capacity Manual* [Transportation Research Board 1985] for the analysis of two-lane highways in urban areas. Ideal conditions for rural two-lane highways are [ibid]:

1.   Design speed greater than or equal to 60 mph.

2.   Lane widths greater than or equal to 12 ft.

3.   Clear shoulders wider than or equal to 6 ft.

4.   No no-passing zones on the highway segment.

TABLE 6.10
Level-of-Service Criteria for Two-Lane Highways

		v / c Ratio[a]																		
		Level Terrain							Rolling Terrain							Mountainous Terrain						
	Percent Time Delay	Avg[b] Speed	Percent No-Passing Zones						Avg[b] Speed	Percent No-Passing zones						Avg[b] Speed	Percent No-Passing Zones					
LOS			0	20	40	60	80	100		0	20	40	60	80	100		0	20	40	60	80	100
A	≤ 30	≥ 58	0.15	0.12	0.09	0.07	0.05	0.04	≥ 57	0.15	0.10	0.07	0.05	0.04	0.03	≥ 56	0.14	0.09	0.07	0.04	0.02	0.01
B	≤ 45	≥ 55	0.27	0.24	0.21	0.19	0.17	0.16	≥ 54	0.26	0.23	0.19	0.17	0.15	0.13	≥ 54	0.25	0.20	0.16	0.13	0.12	0.10
C	≤ 60	≥ 52	0.43	0.39	0.36	0.34	0.33	0.32	≥ 51	0.42	0.39	0.35	0.32	0.30	0.28	≥ 49	0.39	0.33	0.28	0.23	0.20	0.16
D	≤ 75	≥ 50	0.64	0.62	0.60	0.59	0.58	0.57	≥ 49	0.62	0.57	0.52	0.48	0.46	0.43	≥ 45	0.58	0.50	0.45	0.40	0.37	0.33
E	> 75	≥ 45	1.00	1.00	1.00	1.00	1.00	1.00	≥ 40	0.97	0.94	0.92	0.91	0.90	0.90	≥ 35	0.91	0.87	0.84	0.82	0.80	0.78
F	100	< 45	—	—	—	—	—	—	< 40	—	—	—	—	—	—	< 35	—	—	—	—	—	—

Source: Transportation Research Board, *Highway Capacity Manual*, Special Report 209, National Research Council, Washington, D.C., 1985.

[a] Ratio of flow rate to an ideal capacity of 2800 pcph in both directions.

[b] Average travel speed of all vehicles (in mph) for highways with design speed ≥ 60 mph; for highways with lower design speeds, reduce speed by 4 mph for each 10-mph reduction in design speed below 60 mph; assumes that speed is not restricted to lower values by regulation.

5. All passenger cars in the traffic stream.

6. A 50/50 directional split of traffic.

7. No impediments to through traffic due to traffic control or turning vehicles.

8. Level terrain.

The capacity under such conditions is 2800 pcph, total, in both directions. This leads to the basic service flow expression for two-lane, two-way rural highways:

$$SF_i = 2800 \times (v/c)_i \times f_d \times f_w \times f_{HV} \qquad (6.9)$$

where all terms are as defined for freeways and multilane highways with the exception of f_d, which is an additional adjustment factor for the nonideal directional distribution of traffic (i.e., having more than 50 percent of the total traffic traveling in one of the two directions). The logic for f_d is that as the directional distribution of traffic deviates from 50/50 total capacity of the combination of both directions declines until 100 percent of the traffic is traveling in one direction, in which case the capacity of the road reduces to 2000 pcph, which is the equivalent of the 2000 pcphpl used for 60- and 70-mph design speeds for freeways and multilane highways. Values for $(v/c)_i$, f_d, and f_w are obtained from Tables 6.10, 6.11, and 6.12, respectively. The passenger equivalency factors needed to determine f_{HV} (via Eq. 6.7) are obtained from Table 6.13 (with grade steepness and distance criteria satisfied as discussed in Section 6.4.3).

Two points relating to these tables are worthy of note. First, the v/c terms included in Table 6.10, for various levels of service, differ from those given in similar tables for freeways and multilane highways (i.e., Tables 6.1 and 6.5). The v/c terms in Table 6.10 are adjusted to implicitly include reductions in capacity resulting from the combined effects of different terrain types and different percentages of no-passing zones. This simplifies computational procedures (as will be shown in examples), since all volume-to-capacity ratios are given in terms of the ideal capacity of 2800 pcphpl, total, both directions. The second point relates to the adjustment factor f_w (Table 6.12) and the passenger-car equivalency factors (Table 6.13) [Werner and Morrall 1976]. Research has found that these

TABLE 6.11
Adjustment for Directional Distribution on Two-Lane Highways

Directional distribution	100 / 0	90 / 10	80 / 20	70 / 30	60 / 40	50 / 50
Adjustment factor, f_d	0.71	0.75	0.83	0.89	0.94	1.00

Source: Transportation Research Board, *Highway Capacity Manual*, Special Report 209, National Research Council, Washington, D.C., 1985.

TABLE 6.12
Adjustment for the Effects of Narrow Lanes and Restricted Shoulder Width, f_w

Usable[a] Shoulder Width (ft)	12-ft Lanes		11-ft Lanes		10-ft Lanes		9-ft Lanes	
	LOS A–D	LOS[b] E	LOS A–D	LOS[b] E	LOS A–D	LOS[b] E	LOS A–D	LOS[b] E
≥ 6	1.00	1.00	0.93	0.94	0.84	0.87	0.70	0.76
4	0.92	0.97	0.85	0.92	0.77	0.85	0.65	0.74
2	0.81	0.93	0.75	0.88	0.68	0.81	0.57	0.70
0	0.70	0.88	0.65	0.82	0.58	0.75	0.49	0.66

Source: Transportation Research Board, *Highway Capacity Manual*, Special Report 209, National Research Council, Washington, D.C., 1985.
[a] Where shoulder width is different on each side of the roadway, use the average shoulder width.
[b] Factor applies for all speeds less than 45 mph.

TABLE 6.13
Passenger Car Equivalents for Two-Lane Highways

Vehicle Type	Level of Service	Type of Terrain		
		Level	Rolling	Mountainous
Trucks, E_T	A	2.0	4.0	7.0
	B and C	2.2	5.0	10.0
	D and E	2.0	5.0	12.0
Recreational vehicles, E_R	A	2.2	3.2	5.0
	B and C	2.5	3.9	5.2
	D and E	1.6	3.3	5.2
Buses, E_B	A	1.8	3.0	5.7
	B and C	2.0	3.4	6.0
	D and E	1.6	2.9	6.5

Source: A. Werner and J. F. Morrall, "Passenger Car Equivalencies of Trucks, Buses, and Recreational Vehicles for Two-Lane Rural Highways," *Transportation Research Record 615*, 1976.

factors vary by level of service, which was not the case for freeways and multilane highways. As will be shown in forthcoming examples, this dependence on level of service will complicate the traffic analysis procedure.

As a final observation, note that Eq. 6.9 does not contain an adjustment factor for regular/nonregular users as was the case in Eqs. 6.5 and 6.8 (for freeways and multilane highways, respectively). Because of the many other complexities of two-lane highways, the type of driving population has been found not to be a critical concern [Transportation Research Board 1985].

Example 6.5

A rural two-lane highway is on level terrain with 11-ft lanes, 2-ft paved shoulders, and 80 percent no-passing zones. The directional split is 80/20 and there are 5 percent trucks, 2 percent buses, and 5 percent recreational vehicles. Determine the service flow of the roadway at capacity.

Solution

The roadway reaches capacity of LOS E which, from Table 6.10 gives a v/c ratio of 1.0 on level terrain with 80 percent no-passing zones. Thus the service flow at capacity can be computed using Eq. 6.9,

$$SF_E = 2800 \times v/c_E \times f_d \times f_w \times f_{HV}$$

where $v/c_E = 1.0$ (Table 6.10)
$\quad f_d = 0.83$ (80/20 directional split, Table 6.11)
$\quad f_w = 0.88$ (11-ft lanes, 2-ft shoulders, LOS E, Table 6.12)
$\quad E_T = 2.0$, $E_R = 1.6$, $E_B = 1.6$ (level terrain, LOS E, Table 6.13)

From Eq. 6.7, we obtain

$$f_{HV} = \frac{1}{1 + 0.05(2 - 1) + 0.05(1.6 - 1) + 0.02(1.6 - 1)}$$

$$= 0.916$$

Substituting these terms into the above equation for SF_E,

$$SF_E = 2800 \times 1.0 \times 0.83 \times 0.88 \times 0.916$$

$$= \underline{1873.33 \text{ veh/hr}}$$

Example 6.6

Consider the conditions described in Example 6.5. If the peak-hour vehicle count is 522 with a peak-hour factor of 0.90, determine the level of service.

Solution

Note that both f_w and f_{HV} are dependent on the LOS that is not known. Therefore, this problem must be approached by initially assuming a LOS, then computing a LOS, and making certain that the computed LOS is consistent with the initially assumed LOS. To begin, assume LOS E. Under this assumption, f_d, f_w and f_{HV} are all as determined in Example 6.5. Also, from given information,

$$SF = \frac{V}{PHF}$$

$$= \frac{522}{0.90}$$

$$= 580 \text{ veh/hr}$$

To compute LOS (via v/c), Eq. 6.9 is rearranged,

$$v/c = \frac{SF}{2800 \times f_d \times f_w \times f_{HV}}$$

Substituting, we obtain

$$v/c = \frac{580}{2800 \times 0.83 \times 0.88 \times 0.916}$$

$$= \underline{\underline{0.31}}$$

From Table 6.10 (level terrain, 80 percent no-passing zones) LOS C is obtained (i.e., $0.17 < 0.31 < 0.33$), which is inconsistent with the earlier assumed LOS E. Therefore, the above computations must be reworked. If LOS D is assumed, only f_w will change from the assumed LOS E adjustment factors. In this case, $f_w = 0.75$ assuming LOS D with 11-ft lanes and 2-ft shoulders, as indicated in Table 6.12. The v/c is computed as

$$v/c = \frac{580}{2800 \times 0.83 \times 0.75 \times 0.916}$$

$$= \underline{\underline{0.363}}$$

From Table 6.10, LOS D is obtained (i.e., $0.33 < 0.363 < 0.58$), which is consistent with the assumed LOS D. Therefore, the roadways's level of service is D. As a final point, note that if LOS C had been initially assumed, $f_w = 0.75$ (as for the LOS D assumption) but a different f_{HV} would result because now $E_T = 2.2$,

$E_B = 2.0$, and $E_R = 2.5$ (LOS C, level terrain, from Table 6.13). So,

$$f_{HV} = \frac{1}{1 + 0.05(2.2 - 1) + 0.05(2.5 - 1) + 0.02(2.0 - 1)}$$
$$= 0.866$$

By substitution, we find that

$$v/c = \frac{580}{2800 \times 0.83 \times 0.73 \times 0.866}$$
$$= 0.384$$

From Table 6.10 LOS D is again indicated (i.e., $0.33 < 0.384 < 0.58$), which is inconsistent with the assumed LOS C. Thus LOS D provides the only consistent answer.

6.7 DESIGN TRAFFIC VOLUMES

In the preceding sections of this chapter, consideration was given to some predetermined, congested, or peak hour, and the temporal nonuniformity of traffic flow over this hour was accounted for by using the peak-hour factor. However, there is a larger question that looms: How is the peak hour determined for either highway design (i.e., determining the number of lanes required, and so on) or congestion analysis?

This question is complicated by two concerns. First, there is considerable variability in traffic volumes by time of day, day of week, time of year, and type of highway. Figure 6.4 demonstrates such variations in traffic volumes by hour of day, day of week, and type of road. Figure 6.5 gives variations by time of year by comparing flows with the annual average daily traffic AADT (in units of vehicles per day and computed as the total yearly traffic divided by the number of days in the year) [Mutanyi 1963]. Both of these figures underscore the variability in traffic volumes over time. The second concern is an outgrowth of the first in that, given the temporal variability in traffic flow, what hourly volume should be used for design and/or analysis? To answer this question, consider a diagram of the highest 100 hourly volumes on a roadway over a one-year time period, as illustrated in Fig. 6.6. If sufficient capacity is provided for the highest annual hourly volume, the roadway will operate at well below capacity for much of the year. This would be a wasteful use of resources, since additional lanes would be provided for a relatively rare occurrence. In contrast, if the 100th highest volume

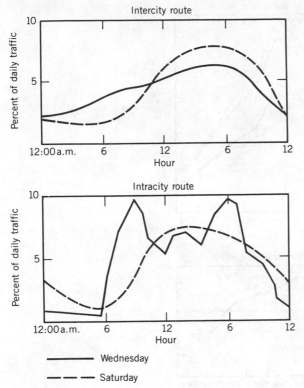

FIGURE 6.4
Examples of hourly variations for inter- and intra-city routes.

is used as capacity for design, the capacity of the roadway will be exceeded 100 times a year, which will result in considerable driver delay. Clearly, some compromise between the expense of providing additional capacity and the expense of incurring additional driver delay must be made.

Current design practice in the United States generally uses a peak hour between the 10th and 50th highest volume hour of the year, depending on the type and location of the road (e.g., urban freeway, rural multilane highway, and so on), local traffic data, and engineering judgment. Perhaps the most common hourly volume used for roadway design is the 30th highest hourly volume of the year. In practice, the K-factor is used to convert average annual daily traffic (AADT) to the 30th highest annual hourly volume. K is defined as

$$K = \frac{\text{DHV}}{\text{AADT}} \tag{6.10}$$

where DHV is the design hourly volume (typically, the 30th highest annual

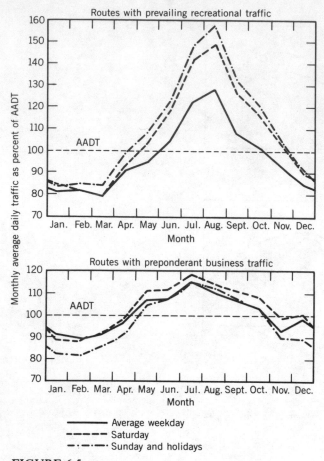

FIGURE 6.5

Examples of monthly traffic volume variations showing relative traffic trends by route type on rural roads (*Source:* T. Mutanyi, "A Method of Estimating Traffic Behavior on All Routes in a Metropolitan County", *Highway Research Record* 41, 1963).

hourly volume) and AADT is the roadway's average annual daily traffic. So, for example, Fig. 6.6 indicates a K value of 0.12. More generally, K_i can be defined as the K-factor corresponding to the ith highest annual hourly volume. Again, for example, the 20th highest annual hourly volume would have a K value, $K_{20} = 0.126$, from Fig. 6.6. If K is not subscripted, the 30th highest annual hourly volume is assumed (i.e., $K = K_{30}$).

Finally, in the design and analysis of some highway types (e.g., freeways and multilane highways), the concern lies with directional traffic flows. Thus a factor is needed to reflect the proportion of peak-hour traffic volume traveling in the peak direction. This factor is denoted as D and is used to arrive at the

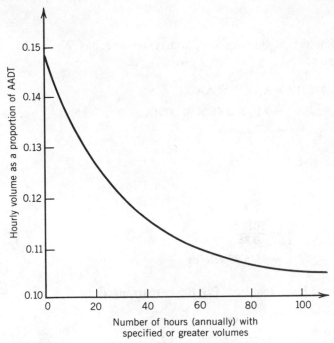

FIGURE 6.6
Highest 100 hourly volumes over a one-year period for a typical roadway.

Directional Design Hour Volume (DDHV) by application of

$$DDHV = K \times D \times AADT \tag{6.11}$$

where all terms are as previously defined.

Example 6.7

A freeway is to be designed as a passenger-car-only facility for an AADT of 31,000 vehicles per day, and 12-ft lanes and a 70-mph design speed are to be used. The design is to be for commuters and the peak-hour factor is 0.85 with 65 percent of the peak-hour traffic traveling in the peak direction. Using Fig. 6.6 to obtain design hour volumes, we determine the number of lanes required (assuming no obstructions) to provide at least LOS C using; the highest annual hourly volume and the 30th highest annual volume.

Solution

By inspection of Fig. 6.6, the highest annual hourly volume has $K_1 = 0.15$. Application of Eq. 6.11 gives

$$DDHV = K_1 \times D \times AADT$$
$$= 0.15 \times 0.65 \times 31,000$$
$$= 3022.5 \text{ veh/hr}$$

With $V = DDHV$, Eq. 6.2 gives,

$$SF = \frac{V}{PHF}$$
$$= \frac{3022.5}{0.85}$$
$$= 3555.88 \text{ veh/hr}$$

To determine the number of lanes required, Eq. 6.6 is rearranged as

$$N = \frac{SF_i}{c_j \times (v/c)_i \times f_w \times f_{HV} \times f_p}$$

where $c_j = 2000$ pcphpl (70-mph design speed)
  $f_{HV} = 1.0$ (with no heavy vehicles)
  $f_w = 1.0$ (12 ft lanes, no obstructions, Table 6.2)
  $f_p = 1.0$ (commuters only, Table 6.4)

For worst LOS C conditions, $v/c_C = 0.77$. Substituting, we obtain

$$N = \frac{3555.58}{2000 \times 0.77 \times 1.0 \times 1.0 \times 1.0}$$
$$= 2.31$$

Since 2.31 lanes are needed, 3 lanes must be provided to achieve at least a LOS C. For the 30th highest annual volume, Fig. 6.6 gives $K_{30} = K = 0.12$, which, when used in Eq. 6.11 gives

$$DDHV = K \times D \times AADT$$
$$= 0.12 \times 0.65 \times 31,000 = 2418 \text{ veh/hr}$$

or

$$SF = \frac{2418}{0.85}$$
$$= 2844.7 \text{ veh/hr}$$

With all other terms as before, the number of lanes needed is

$$N = \frac{2844.7}{2000 \times 0.77 \times 1.0 \times 1.0 \times 1.0}$$
$$= 1.85$$

So, for the 30th highest annual hourly volume, only two lanes are needed to provide LOS C or better as opposed to the three lanes needed to satisfy the level-of-service conditions for the highest annual hourly volume.

NOMENCLATURE
FOR
CHAPTER 6

AADT   average annual daily traffic (in veh/day)

c roadway capacity (in veh/hr)

c_j capacity per lane at design speed j (in pcphpl)

D directional factor

DHV   design-hour volume

DDHV   directional design hour volume

E_B passenger-car equivalent for buses

E_R passenger-car equivalent for recreational vehicles

E_T passenger-car equivalent for trucks

f_d adjustment factor for directional distribution of traffic (two-lane highways only)

f_E adjustment factor for highway type and development environment (multilane highways only)

f_{HV} adjustment factor for heavy vehicles

f_p adjustment factor for driver population (freeways and multilane highways only)

f_w adjustment factor for lane widths and lateral clearances

K factor used to convert AADT to 30th highest annual hourly volume

K_i factor used to convert AADT to the ith highest annual hourly volume

MSF_i maximum service flow rate for level of service i (in pcphpl)

N number of lanes in one direction

PHF   peak-hour factor

SF   service flow (in veh/hr)

V hourly volume (in veh/hr)

V_{15} highest 15 minute volume

v/c   volume-to-capacity ratio

REFERENCES

1. Transportation Research Board, *Highway Capacity Manual*, Special Report 209, National Research Council, Washington, D.C., 1985.

2. Highway Research Board, *Highway Capacity Manual*, Special Report 87, National Research Council, Washington, D.C., 1965.

3. A. Werner and J. F. Morrall, "Passenger Car Equivalencies of Trucks, Buses, and Recreational Vehicles for Two-Lane Rural Highways," *Transportation Research Record 615*, 1976.

4. T. Mutanyi, "A Method of Estimating Traffic Behavior on All Routes in a Metropolitan County," *Highway Research Record 41*, 1963.

PROBLEMS

6.1. A six-lane freeway is operating under the maximum of LOS C conditions, has a design speed of 70 mph, and 10-ft lanes, with 8-ft shoulders on the median side and 4-ft shoulders on the nonmedian side (i.e., roadside). The freeway is on rolling terrain with a commuting traffic composed of 10 percent trucks, 5 percent buses, 85 percent cars, and a PHF = 0.80. If 25 percent of all directional traffic occurs during the peak hour, determine the total daily directional traffic volume.

6.2. A four-lane freeway (two lanes in each direction) is located on rolling terrain and has a design speed of 70 mph, 12-ft lanes, and no lateral obstructions within 8 ft of the pavement edges. The traffic stream consists of cars and trucks only (no buses or recreational vehicles). A weekday peak-hour volume of 1800 vehicles is observed with 500 arriving in the most congested 15-min period. If a level of service no worse than C is desired, determine the maximum number of trucks that can be present in the peak-hour traffic volume.

6.3. Consider the initial conditions described in Example 6.1. How many additional trucks can be added to the peak hour before the freeway reaches capacity?

6.4. A six-lane freeway (three lanes in each direction) has a weekday peak-hour volume of 2400 vehicles (peak-hour factor = 0.90) with 10 percent trucks, 5 percent buses, and 2 percent recreational vehicles. The design speed is 70 mph and the lanes are 11 ft with obstructions on one side (4 ft from the traveled pavement). If the highway is operating at level of service C, what type of terrain is it on?

6.5. What is the maximum directional service flow for a multilane (four lanes, two lanes in each direction) undivided highway at LOS C with 12-ft lanes, a 70-mph design speed, in a rural area with commuting traffic and

    (a) No additional features that limit capacity.

    (b) Telephone poles located 4 ft from pavement edges.

    (c) 10 percent trucks and 5 percent buses on level ground.

    (d) Road located in national park (i.e., noncommuters).

6.6. A level suburban six-lane (three lanes in each direction) undivided multilane facility has a 50-mph design speed with 10 percent trucks, 5 percent buses, 11-ft lanes with obstructions 4 ft from the pavement edge (both sides), and a commuter driving population. The highway currently operates at capacity with a directional peak-hour volume of 2936 veh/hr. If the facility is upgraded to a freeway (70-mph design speed) with a 30-ft median (with no obstructions on the median side but still having obstructions 4 ft from the outside traveled edge of the roadway), what will be the maximum directional peak-hour volume under LOS D? (Assume that *all* lane and traffic related characteristics are unchanged.)

6.7. The new freeway in Problem 6.6 attracts 300 more veh/hr (same PHF). You are asked by an engineering firm to suggest some road improvement that will ensure that the new facility operates under at least level of Service C.

6.8. Consider the conditions described in Example 6.3. How many additional vehicles can be added to the peak hour before the highway reaches capacity?

6.9. A rural divided multilane highway (four lanes, two lanes in each direction) is being designed with 10-ft lanes and a 60-mph design speed. The design traffic consists of regular users and, for the peak hour, is 950 vehicles with 15 percent trucks (no buses or recreational vehicles) and a peak-hour factor of 0.85. The terrain is mountainous and obstructions (rock embankments) are to be cleared from both sides. How much distance must be cleared from the edge of the traveled way to provide a level of service C?

6.10. A multilane highway has a 50-mph design speed and during the peak hour has 1500 vehicles with a peak-hour factor of 0.9. The highway is operating at level of service C with $v/c = 0.48$. How many additional vehicles can be added to the peak hour before the level of service drops to D?

6.11. Consider the conditions described in Example 6.5. If the peak-hour vehicle count is 486 with a peak-hour factor of 0.90, what is the maximum percent of no-passing zones that can exist and still have the highway provide a level of service C (all other physical characteristics as stated in Example 6.5)?

6.12. A rural two-lane highway, with commuters, is currently operating at capacity on mountainous terrain. Passing is not permitted and the road has 11-ft lanes, 2-ft shoulders, (i.e., obstructions 2-ft from the traveled pavement) with a traffic mix of cars and trucks and a 70/30 directional split. A recent traffic count revealed 256 vehicles (total, both directions) arriving in the most congested 15-min interval. The road is widened to a four-lane (two lanes in each direction) multilane divided facility with 60-mph design speed, same lane widths, and 2-ft shoulders on both sides. The widened road attracts additional traffic such that 500 vehicles arrive (total, both directions, same directional split) in the most congested 15-min interval. Determine the new facility's level of service.

6.13. A two-lane rural highway carries a peak-hour volume of 180 veh/hr, and has a PHF = 0.87. The roadway has a 60-mph design speed, 11-ft lanes, 2-ft shoulders (i.e., obstructions 2 ft from the traveled pavement), and is on mountainous terrain with 80 percent no-passing zones. The traffic stream has a 60/40 directional split with 5 percent trucks, 10 percent recreational vehicles, no buses, and 85 percent passenger cars. Determine the level of service.

6.14. A four-lane urban freeway (two lanes in each direction) has a 70-mph design speed, is on mountainous terrain, and has 12-ft lanes with only roadside obstructions (i.e., no median-side obstructions) 4 ft from the traveled pavement. The traffic stream consists of commuters and has 5 percent trucks, 4 percent buses, and 4 percent recreational vehicles. Computations show that the existing volume-to-capacity ratio is 0.82. If the traffic volume remains constant, what percent decrease in the peak-hour factor can be accommodated before the roadway reaches capacity?

6.15. A suburban undivided highway (two lanes in each direction) has a traffic stream of regular users and a 50-mph design speed, 11-ft lanes, obstructions 2 ft from the roadside edge only (i.e., no median-side obstructions), and a peak-hour factor of 0.85. Currently, trucks are not permitted on the highway, but the traffic stream does have buses (10 percent) and recreational vehicles (2 percent) and is on level terrain. The highway is operating (during the peak hour) at maximum LOS C conditions. If trucks are to be permitted to use the highway, determine the maximum number of trucks that can be added to the existing traffic before LOS E is reached.

6.16. An engineer designs a rural two-lane, two-way road on rolling terrain with 10-ft lanes and 2-ft shoulders (i.e., obstructions 2 ft from the traveled pavement). The traffic count is 280 veh/hr (total, both directions) with a peak-hour factor of 0.80 (10 percent trucks, 2 percent buses, and 5 percent recreational vehicles) and there are 40 percent no-passing zones. What directional split is necessary to ensure that the roadway operates at LOS C or better?

6.17. A freeway has 12-ft lanes, no lateral obstructions, passenger cars only, a commuter driving population, a 70-mph design speed, two lanes in each direction, a peak-hour directional distribution of 0.70, and PHF = 0.80. Assuming that Fig. 6.6 applies, if the AADT is 40,000 veh/day, determine the level of service for the 10th, 50th, and 100th highest annual hourly volume.

Chapter Seven

Principles of Traffic Forecasting

7.1 INTRODUCTION

The modification of a highway network either by new road construction or operating improvements on existing roads (e.g., use or retiming of traffic signalization) must be predicated upon some expectation or forecast of roadway traffic volumes. For new road construction, forecasts of traffic are needed to determine an appropriate pavement (number of equivalent axle loads as discussed in Chapter 4) as well as an appropriate geometric design (number of lanes, shoulder widths, and so on) that will provide an acceptable roadway level of service. For operational improvements, traffic forecasts are needed to estimate the effectiveness of alternate improvement options.

In forecasting vehicular traffic, two interrelated elements must be considered: (1) overall regional traffic growth/decline and (2) traffic diversion. The long-term trend of traffic growth/decline is clearly an important concern, since projects such as roadway construction and even operational improvements must be undertaken with some idea of what future traffic conditions will be. In the design of these projects, the engineer must seek to provide a sufficient roadway level of service and an acceptable pavement ride quality for future traffic volumes. One would expect that factors affecting long-term regional traffic growth/decline trends are primarily economic and, to an historically lesser extent, social in nature. The economics of the region in which the roadway action is being undertaken determine the amount of traffic generating activities (e.g., work and shopping) and the spatial distribution of residential, industrial, and commercial areas. The social aspects of the population determine attitudes and behavioral tendencies with regard to traffic-generating decisions. For example, some regional populations may have social characteristics that make them more likely, than other regional populations to make fewer trips, to carpool, to vanpool, or to take public transportation (buses or subways), all of which significantly impact the actual amount of roadway traffic.

In addition to overall regional traffic growth/decline, there is the more microscopic, short-term phenomenon of traffic diversion. As new roads are constructed, as operational improvements are made, and/or as roads gradually become more congested, traffic will divert as drivers change routes or trip-departure times in an effort to avoid congestion and improve the level of service that they experience. Thus the highway network must be viewed as system and, consequently, with the realization that a capacity or level-of-service change on any one roadway of the highway network will impact traffic flows on many of the surrounding roadways.

From the above discussion, it is obvious that traffic forecasting presents itself as a formidable problem, since it requires accurate regional economic forecasts as well as accurate forecasts of trip makers' social and behavioral attitudes regarding trip-oriented decisions (to predict growth/decline trends and traffic diversion). Virtually everyone is aware of how poorly economic forecasts can be, and this is testament to the complexity and uncertainty associated with such forecasts. Similarly, one can readily imagine the difficulty associated with forecasting individuals' trip-oriented decisions.

In spite of the enormous obstacles facing the accurate forecasting of traffic, over the years traffic engineers and analysts have persisted in the development and refinement of a wide variety of traffic forecasting techniques. While, theoretically, traffic forecasting techniques have improved over time, an overriding consideration has always been the ease with which a technique can be implemented in terms of input data requirements and the ability of users to comprehend the underlying methodological approach. The field has evolved such that many traffic analysts can legitimately argue that the more recent developments in traffic forecasting are largely beyond the reach of practice-oriented implementation. In many respects, this is an expected evolution since, due to the complexity of the traffic forecasting problem, there will always be a tendency for theoretical work to exceed the limits of practical implementability. Unfortunately, the outgrowth of the methodological gap between theory and practice has resulted in the use of a wide variety of traffic forecasting techniques, the selection of which is a function of the technical expertise of forecasting agencies' personnel as well as time and financial concerns.

In the past, most textbooks have attempted to cover the full range of traffic forecasting techniques from the readily implementable simplistic approaches to the more theoretically refined methods. In so doing, such textbooks often sacrificed depth of coverage and, as a consequence, traffic forecasting frequently gave the appearance of being confusing and disjoint. This book attempts merely to convey the basic principles underlying traffic forecasting as opposed to reviewing the many techniques available to forecast traffic. This is achieved by focusing on an approach to traffic forecasting that is fairly advanced, technically, and one that, we believe, effectively and efficiently conveys the fundamental concepts of traffic forecasting. For more information on traffic forecasting techniques that are more implementable or more theoretically advanced than the one provided in this chapter, the reader is referred to other sources [Meyer and Miller 1984; Papacostas 1987; Haefner 1986; Morlok 1978; Sheffi 1985; Ben-Akiva and Lerman 1985; Mahmassani and Chang 1987; Abu-Eisheh and Mannering 1988; and Mannering 1989].

7.2 TRAVELER DECISIONS

Forecasts of highway traffic should, at least in theory, be predicated upon some understanding of traveler decisions, since the various decisions that travelers make regarding trips will ultimately determine the quantity, spatial distribution (by route), and temporal distribution (by time) of vehicles on a highway network. Within this context, travelers can be viewed as making four distinct but interrelated decisions regarding trips: (1) temporal decisions, (2) destination decisions,

(3) modal decisions, and (4) spatial or route decisions. The temporal decision includes the decision to travel and, more important, when to travel. The destination decision is concerned with the selection of a specific destination (e.g., shopping center, etc.) while the modal decision relates to how the trip is to be made (e.g., by automobile, bus, walking, or bicycling). Finally, spatial decisions focus on which route is to be taken from the traveler's origin (i.e., the traveler's initial location) to the traveler's desired destination. As can be imagined, being able to understand, let alone predict, such decisions is a monumental task. The remaining sections of this chapter seek to define the dimensions of this decision prediction task and, through examples and illustrations, to demonstrate methods of forecasting traveler decisions and ultimately traffic volumes.

7.3 SCOPE OF THE TRAFFIC FORECASTING PROBLEM

Since traffic forecasting is predicated upon the accurate forecasting of traveler decisions, two factors must be addressed in the development of an effective traffic forecasting methodology: (1) the complexity of the traveler decision-making process and (2) system equilibration.

To begin the development of a fuller understanding of the complexity of traveler decisions (and traffic forecasting), consider the schematic presented in Fig. 7.1. This figure indicates that traveler socioeconomics and activity patterns constitute a major driving force in the decision-making process. Socioeconomics, including factors such as household income, number of household members and traveler age, affect the types of activities that the traveler is likely to be involved in (e.g., work, yoga classes, shopping, children's day care, dancing lessons, community meetings, and so on), which in turn act as a primary factor in many travel decisions. Socioeconomics can also have a direct effect on travel-related decisions by, for example, limiting modal availability (e.g., travelers in low-income households may be forced to take a bus due to the nonavailability of a household automobile).

If we look more directly at the decision to travel, mode/destination choice, and highway route choice, Fig. 7.1 indicates that both long-term and short-term factors affect these decisions. For the decision to travel, as well as mode/destination choice, the long-term factors of modal availability, residential and commercial distributions, and modal infrastructure play a significant role. These factors are considered long-term because they change relatively slowly over time. For example, the development and/or relocation of residential neighborhoods and commercial centers is a process that may take years. Similarly, changes in modal infrastructure (e.g., construction/relocation of highways, subways, commuter rail systems) and modal availability (e.g., changes in automobile ownership, bus routing/scheduling) are also factors that evolve over relatively long periods of time. In contrast, a short-term factor, such as modal traffic, is one that can vary in short periods of time as shown in Chapter 6 (see Fig. 6.1).

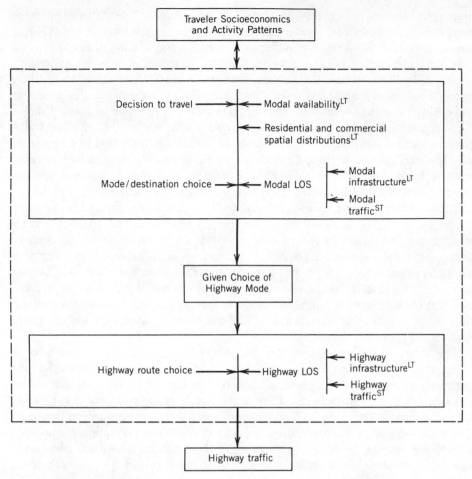

LT: Long-term factors.
ST: Short-term factors.

FIGURE 7.1
Overview of the process by which highway traffic is determined.

Moving further down the illustration presented in Fig. 7.1, we see that the traveler's highway route choice decision is again determined by both long-term (highway infrastructure) and short-term (highway traffic) factors. The outcome of the combination of these traveler decisions is, of course, highway traffic, the prediction of which is the objective traffic forecasting.

Aside from the complexities involved in the traveler decision-making process, the issue of system equilibration (mentioned in the beginning of this section) must also be considered. Note that Fig. 7.1 not only indicates that long- and short-term factors affect traveler decisions and choices, but also that these decisions and choices in turn affect the long and short-term factors. Such a simultaneous relationship is most apparent when considering the relationship between traveler choices and short-term factors. For example, consider a traveler's choice of highway route. One would expect that a traveler would be more likely to select a route, between an origin and a destination, that provides a shorter travel time. The travel time on various routes will be a function of route distance (i.e., highway infrastructure, long-term) and route traffic (i.e., higher traffic volumes reduce travel speed and increase travel times as discussed in Chapter 5). But travelers' decisions to take specific routes ultimately determine the route traffic upon which their route decisions are based. This interdependence between traveler decisions and modal traffic is schematically presented in Fig. 7.2. In addition to these short-term effects, persistently high traffic volumes may result in a change in the highway infrastructure (e.g., construction of additional lanes and/or new highways to reduce congestion) again resulting in an interdependence. This interdependence creates the problem of equilibration that is common to many modeling applications. Perhaps the most recognizable equilibration problem is the determination of price in a classic model of economic supply and demand for a product. From a modeling perspective, as will be shown, equilibration adds yet another dimension of difficulty to an already complex traffic forecasting problem.

It is safe to say that no existing traffic forecasting methodology has come close to accurately capturing the complexities involved in traveler decisions or to fully addressing the issue of equilibration. However, within rather obvious limitations, the field of traffic forecasting has, over the years, made progress toward more accurately modeling traveler decision complexities and equilibration concerns. This evolution of traffic forecasting methodology has led to the popular approach of viewing traveler decisions as a sequence of three distinct decisions, as shown in Fig. 7.3, the result of which is forecasted traffic flow (a direct outgrowth of the

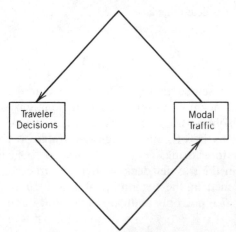

FIGURE 7.2
Interdependence of traveler decisions and traffic flow.

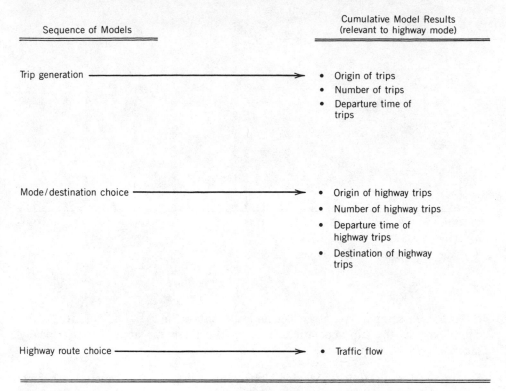

Sequence of Models	Cumulative Model Results (relevant to highway mode)
Trip generation ⟶	• Origin of trips • Number of trips • Departure time of trips
Mode/destination choice ⟶	• Origin of highway trips • Number of highway trips • Departure time of highway trips • Destination of highway trips
Highway route choice ⟶	• Traffic flow

FIGURE 7.3
Overview of the sequential approach to traffic estimation.

highway route choice decision). Clearly, the sequential structure of traveler decisions is a considerable simplification of the actual decision-making process in which all trip-related decisions are considered simultaneously by the traveler. However, this sequential simplification permits the development of a sequence of mathematical models of traveler behavior that can be applied to forecast traffic flow. The following sections of this chapter present and discuss typical functional forms of the mathematical models used to forecast the three sequential traveler decisions shown in Fig. 7.3.

7.4 TRIP GENERATION

The first traveler decision to be modeled in the sequential approach to traffic forecasting is trip generation. The objective of trip generation modeling is to develop an expression that predicts exactly when a trip is to be made. This is an inherently difficult task due to the wide variety of trip types (e.g., working or shopping) and activities (e.g., eating lunch or visiting friends) undertaken by a

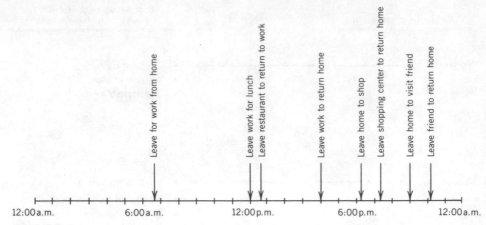

FIGURE 7.4
Weekday trip generation for a typical traveler.

traveler in a sample day, as is schematically shown in Fig. 7.4. To address the complexity of the trip generation decision, the following approach is typically taken:

1. **Aggregation of Decision-Making Units**   Predicting the trip generation behavior is simplified by considering the trip generation behavior of a household (i.e., a group of travelers sharing the same domicile) as opposed to the behavior of individual travelers. Such an aggregation of traveler decisions is justified on the basis of the comparatively homogeneous nature of household members (socially and economically) and their often intertwined trip generating activities (e.g., joint shopping trips, and so on).

2. **Segmentation of Trips by Type**   Different types of trips have different characteristics that make them more or less likely to be taken at various times of the day. For example, work trips are more likely to be taken in the morning hours than are shopping trips, which are more likely to be taken during the evening hours. Also, it is more likely that the traveler will have two or three shopping trips during the course of a day than two or three work trips. To account for this, three distinct trip types are used: (1) work trips, including trips to and from work, (2) shopping trips, and (3) social/recreational trips, which include vacations, visiting friends, church meetings, sporting events, and so on.

3. **Temporal Aggregation**   Although research has been undertaken to develop mathematical expressions that predict when a traveler is likely to make a trip [Mahmassani and Chang 1987; Abu-Eisheh and Mannering 1988; and Mannering 1989], more often, trip generation focuses on the number of trips made over some period of time. Thus trips are aggregated temporally, and trip generation models seek to predict the number of trips per hour or per day.

7.4.1 Typical Trip Generation Models

Trip generation models generally assume a linear form, in which the number of vehicle-based (automobile, bus, or subway) trips is a function of various socio-, economic and/or distributional (residential and commercial) characteristics. An example of such a model is,

$$T_{ij} = b_0 + b_1 z_{1j} + b_2 z_{2j} + \cdots + b_n z_{nj} \tag{7.1}$$

where T_{ij} is the number of vehicle-based trips in some specified time period of type i (i.e., shopping or social/recreational) made by household j, z_{nj} is characteristic n (e.g., income, employment in neighborhood, number of household members) of household j, and b_n is a coefficient estimated from traveler survey data and corresponding to characteristic n.

The estimated coefficients (b's) are most often estimated by the method of least squares regression using data collected from traveler surveys. A brief description and example of this method are presented in Appendix 7A.

Example 7.1

A simple linear regression model is estimated for shopping trip generation during a shopping-trip peak hour (e.g., Saturday afternoon). The model is

> number of peak-hour vehicle-based shopping trips per household
> = 0.12 + 0.09(household size)
> + 0.011 (annual household income (in thousands of dollars))
> − 0.15 (employment in the household's neighborhood (in hundreds))

A particular household has six members, and an annual income of $50,000. They live in a neighborhood with 450 retail employees. They move to a new home in a neighborhood with 150 retail employees. Calculate the predicted number of peak-hour shopping trips the household makes before and after the move.

Solution

Note that the signs of the model coefficients (b's, $+0.09$, and $+0.011$) indicate that as household size and income increase, the number of shopping trips also increase. This is reasonable since wealthier, larger households can be expected to make more vehicle-based shopping trips. The negative sign of the employment coefficient (-0.15) indicates that as retail employment in a household's neighborhood increases, fewer vehicle-based shopping trips will be generated. This reflects the fact that a larger neighborhood retail employment implies more shopping opportunities nearer to the household, thereby increasing the possibility that a shopping trip can be conducted without the use of a vehicle (i.e., a nonvehicle-based trip, such as walking).

Turning to the problem solution, before the household moves, we find that

$$\text{number of trips} = 0.12 + 0.09(6) + 0.011(50) - 0.15(4.5)$$
$$= 0.535 \text{ vehicle trips}$$

After the household moves, we obtain

$$\text{number of trips} = 0.12 + 0.09(6) + 0.011(50) - 0.15(1.5)$$
$$= 0.985 \text{ vehicle trips}$$

Thus the model predicts that the move will result in 0.45 additional peak-hour vehicle-based shopping trips due to the decline in neighborhood shopping opportunities as reflected by the decline in neighborhood retail employment.

Example 7.2

A model for social/recreational trip generation is estimated with data collected during a major holiday, as

number of peak-hour vehicle-based social/recreational trips per household

$$= 0.04 + 0.018 \text{ (household size)}$$
$$+ 0.009 \text{ (annual household income (in thousands of dollars))}$$
$$+ 0.16 \text{ (number of nonworking household members)}$$

If the family described in Example 7.1 has one working member, how many peak-hour social/recreational trips are predicted?

Solution

The positive signs of the model coefficients indicate that increasing household size, income, and number of nonworking household members results in more social/recreational trips. Again, wealthier and larger households can be expected to be involved in more vehicle-based trip generating activities, and the larger the number of nonworking household members, the larger the number of people available, at home, to make peak-hour social/recreational trips.

The solution to this problem simply is

$$\text{number of trips} = 0.04 + 0.018(6) + 0.009(50) + 0.16(5)$$
$$= 1.398 \text{ vehicle trips}$$

Example 7.3

A neighborhood has a retail employment of 205 and 700 households that can be categorized into four types, each type having identical characteristics as follows:

Type	Household Size	Annual Income	Number of Nonworkers in the Peak Hour	Workers Departing
1	2	$40,000	1	1
2	3	$50,000	2	1
3	3	$55,000	1	2
4	4	$40,000	3	1

There are 100 Type 1, 200 Type 2, 350 Type 3, and 50 type 4 households, assuming shopping, social/recreational, and work vehicle-based trips all peak at the same time (for exposition purposes). Determine the total number of peak hour trips (work, shopping, social/recreational) using the generation models described in examples 7.1 and 7.2.

Solution

For vehicle-based shopping trips, we obtain

Type 1: $0.12 + 0.09(2) + 0.011(40) - 0.15(2.05) = 0.4325$ trips/household
$\times$ 100 households
$= 43.25$ trips

Type 2: $0.12 + 0.09(3) + 0.011(50) - 0.15(2.05) = 0.6325$ trips/household
$\times$ 200 households
$= 126.5$ trips

Type 3: $0.12 + 0.09(3) + 0.011(55) - 0.15(2.05) = 0.6875$ trips/household
$\times$ 350 households
$= 240.625$ trips

Type 4: $0.12 + 0.09(4) + 0.011(40) - 0.15(2.05) = 30.625$ trips/household
$\times$ 50 households
$= 30.625$ trips

Therefore, there will be a total of 441 vehicle-based shopping trips.

For vehicle-based social/recreational trips, we obtain

Type 1: $0.04 + 0.018(2) + 0.009(40) + 0.16(1) = 0.596$ trips/household
$$\times \ 100 \text{ households}$$
$$= 59.6 \text{ trips}$$

Type 2: $0.04 + 0.018(3) + 0.009(50) + 0.16(2) = 0.864$ trips/household
$$\times \ 200 \text{ households}$$
$$= 172.8 \text{ trips}$$

Type 3: $0.04 + 0.018(3) + 0.009(55) + 0.16(1) = 0.749$ trips/household
$$\times \ 350 \text{ households}$$
$$= 262.15 \text{ trips}$$

Type 4: $0.04 + 0.018(4) + 0.009(40) + 0.16(3) = 0.952$ trips/household
$$\times \ 50 \text{ households}$$
$$= 47.6 \text{ trips}$$

Therefore, there will be a total of 542.15 vehicle-based social/recreational trips.

For vehicle-based work trips, there will be 100 generated from Type 1 households (1×100), 200 from Type 2 (1×200), 700 from Type 3 (2×350), and 50 from Type 4 (1×50) for a total of 1050 vehicle-based work trips. Summing the totals of all three trip types gives 2033 peak-hour vehicle-based trips.

It should be noted that the trip generation models used in Examples 7.1, 7.2, and 7.3 are a simplified representation of the actual trip generation decision-making process. First, there are many more traveler and household characteristics that affect trip generating behavior (age, lifestyles, and so on), and, second, the models have no variables to capture the equilibration concept discussed earlier. The equilibration concern is important, since if the highway system is heavily congested, travelers are likely to make fewer peak-hour trips either as a result of canceling trips or postponing them until a less congested time period. Unfortunately, such obvious model defects must often be accepted due to data and/or resource limitations. The reader is referred to Meyer and Miller [1984] for further discussion of this point.

7.5 MODE AND DESTINATION CHOICE

After the number of trips generated per unit time is known, the next step in the sequential approach to traffic forecasting is the determination of traveler mode and destination. As was the case with trip generation, trips are classified as work, shopping, and social/recreational. For both shopping and social/recreational trips, the traveler will have the option to choose a mode of travel (e.g., automobile, vanpool, or bus) as well as a destination (e.g., different shopping centers). In contrast, the work trip offers only the mode option, since the choice of work location (destination) is usually a long-term decision that is beyond the time range of most traffic forecasts.

7.5.1 Theoretical Approach

Following recent advances in the traffic forecasting field, the development of a mode/destination choice model necessitates the use of some consistent theory of travelers' decision-making process. Of the decision-making theories available, one that is based on the microeconomic concept of utility maximization has enjoyed comparatively widespread acceptance in mode/destination choice modeling. The basic assumption is that a traveler will select the combination of a mode and destination that provides the most utility in the economic sense. The problem, then, becomes one of developing an expression for the utility provided by various mode/destination alternatives. Since it is unlikely that individual travelers' utility functions can ever be specified with certainty, the unspecifiable portion is assumed to be random. To illustrate this approach, consider a utility function of the following form,

$$V_{mk} = \sum_n b_{mn} z_{kmn} + \varepsilon_{mk} \tag{7.2}$$

where V_{mk} is the total utility (specifiable and unspecifiable) provided by mode/destination alternative m to a traveler k; b_{mn} is the coefficient estimated from traveler survey data for mode/destination alternative m corresponding to mode/destination or traveler characteristic n; z_{kmn} is the traveler or mode/destination characteristic n (e.g., income, travel time of mode, commercial floor space at destination, or population at destination) for mode/destination alternative m for traveler k; and ε_{mk} is the unspecifiable portion of the utility of mode/destination alternative m for traveler k, which will be assumed to be random. For notational convenience, define the specifiable nonrandom portion of utility V_{mk} as

$$U_{mk} = \sum_n b_{mn} z_{kmn} \tag{7.3}$$

With these definitions of utility, the probability that a traveler will choose some alternative, say m, is equal to the probability that alternative's utility is greater than the utility of all other possible alternatives. The probabilistic component arises from the fact that the unspecifiable portion of the utility expression is not known and is assumed to be a random variable. The basic probability statement is

$$P_{mk} = \text{prob}[U_{mk} + \varepsilon_{mk} > U_{sk} + \varepsilon_{sk}] \qquad \text{for all } s \neq m \qquad (7.4)$$

Where P_{mk} is the probability that traveler k will select alternative m, prob$[\cdot]$ is notation for probability, and s is the notation for available alternatives. With this basic probability and utility expression and an assumed random distribution of the unspecifiable components of utility (ε_{mk}), a probabilistic choice model can be derived and the coefficients in the utility function (b_{mn}'s in Eqs. 7.2 and 7.3) can be estimated with data collected from traveler surveys, along the same lines as was done for the coefficients in the trip generation models. A popular approach to deriving such a probabilistic choice model is to assume that the random, unspecifiable component of utility (ε_{mk} in Eq. 7.2) is generalized extreme value distributed. With this assumption, a rather lengthy and involved derivation (see McFadden [1981]) gives rise to the logit model formulation,

$$P_{mk} = \frac{e^{U_{mk}}}{\sum_s e^{U_{sk}}} \qquad (7.5)$$

where e is the base of the natural logarithm (i.e., $e = 2.718$).

The coefficients that comprise the specifiable portion of utility (b_{mn}'s in Eq. 7.3) are estimated by the method of maximum likelihood, which essentially accomplishes the same objective as the least squares estimation procedure described in Appendix 7A. For further information on logit model coefficient estimation and maximum likelihood estimation techniques, the reader is referred to more specialized references [Ben-Akiva and Lerman 1985; Mannering 1989; and McFadden 1981].

7.5.2 Logit Model Applications

With the total number of vehicle-based trips made in specific time periods known (from trip generation models), the allocation of trips to vehicle-based modes and likely destinations can be undertaken by applying appropriate logit models. This process is best demonstrated by example.

Example 7.4

A simple work mode choice model is estimated from data in a small urban area to determine the probabilities of individual travelers selecting various modes. The mode choices include auto-drive alone (DL), auto-shared ride (SR) and bus (B), and the utility functions are estimated as

$$U_{DL} = 2.2 - 0.2(\text{cost}_{DL}) - 0.03(\text{travel time}_{DL})$$
$$U_{SR} = 0.8 - 0.2(\text{cost}_{SR}) - 0.03(\text{travel time}_{SR})$$
$$U_B = -0.2(\text{cost}_B) - 0.01(\text{travel time}_B)$$

where cost is in dollars and time is in minutes. Between a residential area and an industrial complex, 4000 workers (generating vehicle-based trips) depart for work during the peak hour. For all workers, the cost of driving an automobile is $4.00 with a travel time of 20 minutes and the bus fare is 50 cents with a travel time of 25 minutes. If the shared-ride option always consists of two travelers sharing costs equally, how many workers will take each mode?

Solution

Note that the utility function coefficients logically indicate that as modal costs and travel times increase, modal utilities decline and, consequently, so do modal selection probabilities (see Eq. 7.5). Substitution of cost and travel time values into the utility expressions gives

$$U_{DL} = 2.2 - 0.2(4) - 0.03(20)$$
$$= 0.8$$
$$U_{SR} = 0.8 - 0.2(2) - 0.03(20)$$
$$= -0.2$$
$$U_B = -0.2(0.5) - 0.01(25)$$
$$= -0.35$$

Substituting these values into Eq. 7.5 yields

$$P_{DL} = \frac{e^{0.8}}{e^{0.8} + e^{-0.2} + e^{-0.35}}$$
$$= \frac{2.226}{2.226 + 0.819 + 0.705}$$
$$= \frac{2.226}{3.749}$$
$$= 0.594$$
$$P_{SR} = \frac{0.819}{3.749}$$
$$= 0.218$$
$$P_B = \frac{0.705}{3.749}$$
$$= 0.188$$

Multiplying these probabilities by 4000 (the total number of workers departing in the peak hour) gives 2380 workers driving alone, 870 sharing a ride, and 750 using a bus.

Example 7.5

The bus company is making costly efforts in an attempt to increase work-trip bus usage for the travel conditions described in Example 7.4. An exclusive bus lane is constructed that reduces bus travel time to 10 min.
(a) Determine the modal distribution of trips after the lane is constructed.
(b) If shared-ride vehicles are also permitted to use the facility and travel time for bus and shared-ride modes is 10 min, determine the modal distribution.
(c) Given the conditions described in part (b), determine the modal distribution if the bus company offers free bus service.

Solution

(a) After the bus lane construction, the modal utilities of drive alone and shared ride are unchanged from those in Example 7.4. However, the bus modal utility becomes

$$U_B = -0.2(0.5) - 0.01(10)$$
$$= -0.2$$

From Eq. 7.5 with 4000 work trips, we find that

$$P_{DI.} = \frac{e^{0.8}}{e^{0.8} + e^{-0.2} + e^{-0.2}}$$

$$= \frac{2.226}{3.8635}$$

$$= 0.576 \quad \text{and} \quad 0.576(4000) = 2304 \text{ trips}$$

$$P_{SR} = \frac{0.819}{3.8635} = 0.212 \quad \text{and} \quad 0.212(4000) = 848 \text{ trips}$$

$$P_B = \frac{0.819}{3.8635} = 0.212 \quad \text{and} \quad 0.212(4000) = 848 \text{ trips}$$

or an increase of 98 bus patrons, from the prediction of Example 7.4.
(b) With the bus lane opened to shared-ride vehicles, only the modal utility of shared ride will change from those in part (a),

$$U_{SR} = 0.08 - 0.02(2) - 0.03(10)$$
$$= 0.1$$

From Eq. 7.5 with 4000 work trips, we obtain

$$P_{DL} = \frac{e^{0.8}}{e^{0.8} + e^{0.1} + e^{-0.2}}$$

$$= \frac{2.226}{4.15}$$

$$= 0.536; \, 0.536(4000) = 2144 \text{ trips}$$

$$P_{SR} = \frac{1.105}{4.15} = 0.267; \, 0.267(4000) = 1068 \text{ trips}$$

$$P_{B} = \frac{0.8187}{4.15} = 0.197; \, 0.197(4000) = 788 \text{ trips}$$

or a loss of 60 bus patrons and a gain of 220 shared-ride users relative to part (a). (c) With free bus fare, the bus modal utility becomes [with other utilities unchanged from part (b)],

$$U_B = -0.2(0) - 0.01(10)$$

$$= -0.1$$

From Eq. 7.5 with 4000 work trips, we obtain

$$P_{DL} = \frac{e^{0.8}}{e^{0.8} + e^{0.1} + e^{-0.1}}$$

$$= \frac{2.226}{4.236}$$

$$= 0.525; \, 0.525(4000) = 2102 \text{ trips}$$

$$P_{SR} = \frac{1.105}{4.236} = 0.261; \, 0.261(4000) = 1043 \text{ trips}$$

$$P_{B} = \frac{0.905}{4.236} = 0.214; \, 0.214(4000) = 855 \text{ trips}$$

or 67 more bus patrons compared to part (b).

Example 7.6

Consider a residential area and two shopping centers that are possible destinations. From 7:00 to 8:00 p.m. on Friday night, 900 vehicle-based shopping trips leave the residential area for the two shopping centers. A joint shopping trip destination/mode choice logit model (choice of either auto or bus) is estimated giving the following coefficients:

Variable	Auto Coefficient	Bus Coefficient
Auto constant	0.6	0.0
Travel time in minutes	-0.3	-0.3
Commercial floor space (in thousands of square feet)	0.012	0.012

Initial travel times to shopping centers 1 and 2 are

	By Auto	By Bus
Travel time to shopping center 1 (in minutes)	8	14
Travel time to shopping center 2 (in minutes)	15	22

If shopping center 2 has 400,000 sq ft of commercial floor space and shopping center 1 has 250,000 sq ft, determine the distribution of Friday night shopping trips by destination and mode.

Solution

The utility function coefficients indicate that as modal travel times increase, the likelihood of selecting the mode/destination combination declines. Also, as the destination's floor space increases, the probability of selecting that destination will increase as suggested by the positive coefficient ($+0.012$). This reflects the fact that bigger shopping centers will tend to have a greater variety of merchandise and hence be more attractive shopping destinations. Note that since this is a joint mode/destination choice model, there are four mode/destination combinations and four corresponding utility functions. Let U_{A1} be the utility of the auto mode to shopping center 1, U_{A2} be the utility of the auto mode to shopping center 2, and U_{B1} and U_{B2} be the utility of the bus mode to shopping centers 1 and 2, respectively. The utilities are

$$U_{A1} = 0.6 - 0.3(8) + 0.012(250)$$
$$= 1.2$$
$$U_{B1} = -0.3(14) + 0.012(250)$$
$$= -1.2$$
$$U_{A2} = 0.6 - 0.3(15) + 0.012(400)$$
$$= 0.9$$
$$U_{B2} = -0.3(22) + 0.012(400)$$
$$= -1.8$$

Substituting these values into Eq. 7.5 gives

$$P_{A1} = \frac{3.32}{6.246}$$
$$= 0.532$$
$$P_{B1} = \frac{0.301}{6.246}$$
$$= 0.048$$
$$P_{A2} = \frac{2.46}{6.246}$$
$$= 0.394$$
$$P_{B2} = \frac{0.165}{6.246}$$
$$= 0.026$$

Multiplying these probabilities by the 900 trips gives 479 trips by auto to shopping center 1, 43 trips by bus to shopping center 1, 355 trips by auto to shopping center 2, and 23 trips by bus to shopping center 2.

Example 7.7

A joint destination/mode vehicle-based social/recreational trip logit model is estimated with the following coefficients:

Variable	Auto Coefficient	Bus Coefficient
Auto constant	0.9	0.0
Travel time in minutes	−0.22	−0.22
Population in thousands	0.16	0.16
Amusement floor space in thousands of square feet	0.11	0.11

It is known that 500 social/recreational trips will depart from a residential area during the peak hour. There are three possible trip destinations with the following characteristics:

	Travel Time (in minutes)		Population (in thousands)	Amusement Floor Space (in thousands ft^2)
	Auto	Bus		
Destination 1	14	17	12.4	13.0
Destination 2	5	8	8.2	9.2
Destination 3	18	24	5.8	21.0

Determine the distribution of trips by mode and destination.

Solution

As was the case for the shopping mode/destination model presented in Example 7.6, the signs of the coefficient estimates indicate that increasing travel time decreases an alternative's selection probability. Also, increasing population (reflecting an increase in social opportunities) and increasing amusement floor space (reflecting more recreational opportunities) both increase the probability of an alternative being selected. With two modes and three destinations, there are six alternatives providing the following utilities (using the same subscripting notation as in Example 7.6),

$$U_{A1} = 0.9 - 0.22(14) + 0.16(12.4) + 0.11(13)$$
$$= 1.234$$
$$U_{B1} = -0.22(17) + 0.16(12.4) + 0.11(13)$$
$$= -0.326$$
$$U_{A2} = 0.9 - 0.22(5) + 0.16(8.2) + 0.11(9.2)$$
$$= 2.124$$
$$U_{B2} = -0.22(8) + 0.16(8.2) + 0.11(9.2)$$
$$= 0.564$$
$$U_{A3} = 0.9 - 0.22(18) + 0.16(5.8) + 0.11(21)$$
$$= 0.178$$
$$U_{B3} = -0.22(24) + 0.16(5.8) + 0.11(21)$$
$$= -2.042$$

Using Eq. 7.5 with 500 trips, the total number of trips to the six mode/destination alternatives are

$$P_{A1} = \frac{3.435}{15.607}$$
$$= 0.22 \quad \text{and} \quad 0.22 \times 500 = 110 \text{ trips}$$
$$P_{B1} = \frac{0.722}{15.607}$$
$$= 0.046 \quad \text{and} \quad 0.046 \times 500 = 23 \text{ trips}$$
$$P_{A2} = \frac{8.365}{15.607}$$
$$= 0.536 \quad \text{and} \quad 0.536 \times 500 = 268 \text{ trips}$$
$$P_{B2} = \frac{1.76}{15.607}$$
$$= 0.113 \quad \text{and} \quad 0.113 \times 500 = 57 \text{ trips}$$
$$P_{A3} = \frac{1.195}{15.607}$$
$$= 0.077 \quad \text{and} \quad 0.077 \times 500 = 38 \text{ trips}$$
$$P_{B3} = \frac{0.13}{15.607}$$
$$= 0.008 \quad \text{and} \quad 0.008 \times 500 = 4 \text{ trips}$$

Example 7.8

Consider the situation described in Example 7.7. A labor dispute results in a bus union slowdown that increases travel times from the origin by 4, 2, and 8 min to destinations 1, 2, 3, respectively. If the total number of trips remains constant, determine the resulting distribution of trips by mode and destination.

Solution

The mode/destination utilities are computed as

$$U_{A1} = 1.234 \text{ (as in Example 7.7)}$$
$$U_{B1} = -0.22(21) + 0.16(12.4) + 0.11(13)$$
$$= -1.206$$
$$U_{A2} = 2.124 \text{ (as in Example 7.7)}$$
$$U_{B2} = -0.22(10) + 0.16(8.2) + 0.11(9.2)$$
$$= -0.124$$
$$U_{A3} = 0.178 \text{ (as in Example 7.7)}$$
$$U_{B3} = -0.22(32) + 0.16(5.8) + 0.11(21)$$
$$= -3.802$$

Applying Eq. 7.5 with 500 trips gives the following distribution of trips among mode/destination alternatives:

$$P_{A1} = \frac{3.435}{14.45}$$
$$= 0.238 \qquad \text{and} \qquad 0.238 \times 500 = 119 \text{ trips}$$

$$P_{B1} = \frac{0.299}{14.45}$$
$$= 0.021 \qquad \text{and} \qquad 0.021 \times 500 = 10 \text{ trips}$$

$$P_{A2} = \frac{8.365}{14.45}$$
$$= 0.579 \qquad \text{and} \qquad 0.579 \times 500 = 290 \text{ trips}$$

$$P_{B2} = \frac{1.132}{14.45}$$
$$= 0.078 \qquad \text{and} \qquad 0.078 \times 500 = 39 \text{ trips}$$

$$P_{A3} = \frac{1.195}{14.45}$$
$$= 0.083 \qquad \text{and} \qquad 0.083 \times 500 = 41 \text{ trips}$$

$$P_{B3} = \frac{0.022}{14.45}$$
$$= 0.002 \qquad \text{and} \qquad 0.002 \times 500 = 1 \text{ trip}$$

7.6 HIGHWAY ROUTE CHOICE

To summarize, the trip generation and mode/destination choice models give total highway traffic demand between a specified origin (e.g., the neighborhood from which trips originate) and a destination (e.g., the geographic area to which trips are destined), in terms of vehicles per some time period (usually vehicles per hour). With this information in hand, the final step in the sequential approach to traffic forecasting, route choice, can be addressed. The result of the route choice decision will be traffic flow (generally in units of vehicles per hour) on specific highway routes, which is the desired output from the traffic forecasting process.

7.6.1 Highway Performance Functions

Route choice presents itself as a classic equilibrium problem, since travelers' route choice decisions are primarily a function of route travel times that are determined by traffic flow, which is itself a product of route choice decisions. This interrelationship between route choice decisions and traffic flow forms the basis of route choice theory and model development.

To begin modeling traveler route choice, a mathematical relationship between route travel time and route traffic flow is needed. Such a relationship is commonly referred to as a highway performance function. The most simplistic approach to formalizing this relationship is to assume a linear highway performance function in which travel time increases linearly with speed. An example of such a function is illustrated in Fig. 7.5. In this figure, the free-flow travel time

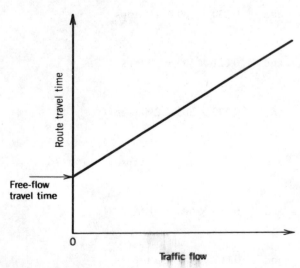

FIGURE 7.5
Linear travel time/flow relationship.

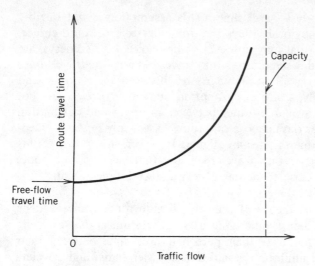

FIGURE 7.6
Nonlinear travel time/flow relationship.

refers to the travel time that a traveler would experience if no other vehicles were present to impede travel speed (as discussed in Chapter 5, Section 5.3.1). This free-flow speed is generally computed assuming that a vehicle travels at the speed limit of the route.

Although the linear highway performance function has the appeal of simplicity, it is not a particularly realistic representation of the travel time/traffic flow relationship. Recall that Chapter 5 (see Fig. 5.3) presented a relationship between traffic speed and flow that was parabolic in nature, with significant reductions in travel speed occurring as the traffic flow approached the roadway's capacity. This parabolic speed/flow relationship suggests a nonlinear highway performance function, such as that illustrated in Fig. 7.6. This figure shows route travel time increasing more quickly as traffic flow approaches capacity, which is consistent with the parabolic relationship presented in Chapter 5.

Both linear and nonlinear highway performance functions will be demonstrated, through example, using two theories of travel route choice; user equilibrium and system optimization (for other theories of route choice, the reader is referred to Sheffi [1985]).

7.6.2 Theory of User Equilibrium

In developing theories of traveler route choice, two important assumptions are usually made. First, it is assumed that travelers will select routes between origins and destinations on the basis of route travel times only (i.e., they will tend to select the route with the shortest travel time). This assumption is not terribly restrictive, since travel time obviously plays the dominant role in route choice, but other more subtle factors that may influence route choice (e.g., scenery), are not accounted for. The second assumption is that travelers know the travel times that would be encountered on all available routes between their origin and destination. This is, potentially, a strong assumption, since the traveler may not have actually traveled on all available routes between an origin and destination and may repeatedly (day after day) choose one route based only on the perception that travel times on alternate routes are higher. However, in support of this assumption, studies have shown that traveler's perceptions of alternate route travel times are reasonably close to actual observed travel times [Mannering 1989].

With these assumptions, the theory of user equilibrium route choice can be operationalized. The rule of choice underlying user equilibrium is that travelers will select a route so as to minimize their personal travel time between their origin and destination. User equilibrium is said to exist when individual travelers cannot improve their travel times by unilaterally changing routes. Stated differently [Wardrop 1952] user equilibrium can be defined as

The travel time between a specified origin and destination on all used routes is equal, and less than or equal to the travel time that would be experienced by a traveler on any unused route.

Example 7.9

Two routes connect a city and a suburb. During the peak-hour morning commute, a total of 4500 vehicles travel from the suburb to the city. Route 1 has a 60-mph speed limit and is 6 miles in length; route 2 is 3 miles in length with a 45-mph speed limit. Studies show that the total travel time on route 1 increases two minutes for every additional 500 vehicles added. Minutes of travel time on route 2 increase with the square of the number of vehicles expressed in thousands of vehicles per hour. Determine user equilibrium travel times.

Solution

Determining free-flow travel times, in minutes, gives

$$\text{Route 1: 6 miles}/60 \text{ mph} \times 60 \text{ min/hr} = 6 \text{ min}$$
$$\text{Route 2: 3 miles}/45 \text{ mph} \times 60 \text{ min/hr} = 4 \text{ min}$$

With these data, the performance functions can be written as

$$t_1 = 6 + 4x_1$$
$$t_2 = 4 + x_2^2$$

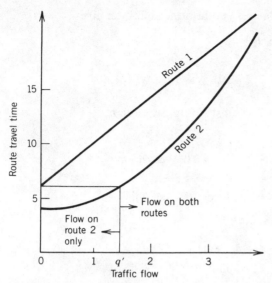

FIGURE 7.7
Illustration of performance curves for Example 7.9.

where t_1 and t_2 are the average travel times on routes 1 and 2 in minutes and x is the traffic flow in thousands of vehicles per hour. Also, we have the basic flow conservation identity,

$$q = x_1 + x_2 = 4.5$$

where q is the total traffic flow between the origin and destination in thousands of vehicles per hour. With Wardrop's definition of user equilibrium, it is known that the travel times on all used routes are equal. However, the first order of concern is to determine whether or not both routes are used. Figure 7.7 gives a graphic representation of the two performance functions. Note that since route 2 has a lower free-flow travel time, any total origin to destination traffic flow less than q' (in Fig. 7.7) will result in only route 2 being used, since the travel time on route 1 would be greater even if only one vehicle used it. At flows of q' and above, route 2 is sufficiently congested, and its travel time sufficiently high, so that route 1 becomes a viable alternative. To check if the problem's flow of 4500 vehicles per hour exceeds q' the following test is conducted:

(a) Assume that all traffic flow is on route 1. Substituting traffic flows of 4.5 and 0 into the performance functions gives $t_1(4.5) = 24$ min and $t_2(0) = 4$ min.

(b) Assume that all traffic flow is on route 2, giving $t_1(0) = 6$ min and $t_2(4.5) = 24.25$ min.

Thus, since $t_1(4.5) > t_2(0)$ and $t_2(4.5) > t_1(0)$, both routes will be used. If $t_1(0)$ would have been greater than $t_2(4.5)$, the 4500 vehicles would have been less than q' in Fig. 7.7 and only route 2 would have been used.

With both routes used, Wardrop's user equilibrium definition gives

$$t_1 = t_2$$

or

$$6 + 4x_1 = 4 + x_2^2$$

From flow conservation, $x_1 + x_2 = 4.5$. Substituting, we obtain

$$6 + 4(4.5 - x_2) = 4 + x_2^2$$
$$x_2 = 2.899 \quad \text{or 2899 veh/hr}$$
$$x_1 = 4.5 - x_2 = 4.5 - 2.899$$
$$= 1.601 \quad \text{or 1601 veh/hr}$$

which gives average route travel times of

$$t_1 = 6 + 4(1.601)$$
$$= 12.4 \text{ min}$$
$$t_2 = 4 + (2.899)^2$$
$$= 12.4 \text{ min}$$

Example 7.10

Peak-hour traffic demand between an origin and destination pair is initially 3500 vehicles. The two routes connecting the pair have performance functions $t_1 = 2 + 3(x_1/c_1)$ and $t_2 = 4 + 2(x_2 c_2)$, where t's are in minutes, and flows (x's) and capacities (c's) are in thousands of vehicles per hour. Initially, the capacities of routes 1 and 2 are 2500 and 4000, respectively. A reconstruction project reduces capacity on route 2 to 2000 veh/hr. Assuming user equilibrium before and during reconstruction, what reduction in total peak-hour origin-destination traffic flow is needed to ensure that total travel times (i.e., summation of all $x_a t_a$'s, where a denotes route) before reconstruction equal those during reconstruction.

Solution

First, focusing on the roads before reconstruction, the check to see if both routes are used gives (using performance functions):

$$t_1(3.5) = 6.2 \text{ min}; \ t_2(0) = 4 \text{ min}$$
$$t_1(0) = 2 \text{ min}; \ t_2(3.5) = 5.75 \text{ min}$$

which, since $t_1(3.5) > t_2(0)$ and $t_2(3.5) > t_1(0)$, indicates that both routes are used. Setting route travel times equal and substituting performance functions gives

$$2 + \frac{3}{2.5}(x_1) = 4 + \frac{2}{4}(x_2)$$

and, from conservation of flow, $x_2 = 3.5 - x_1$, so that

$$2 + 1.2x_1 = 4 + 0.5(3.5 - x_1)$$

Solving gives $x_1 = 2.206$ and $x_2 = 3.5 - 2.206 = 1.294$. For travel times,

$$t_1 = 2 + 1.2(2.206)$$
$$= 4.647 \text{ min}$$
$$t_2 = 4 + 0.5(1.294)$$
$$= 4.647 \text{ min}$$

The total peak-hour travel time before reconstruction will simply be the average route travel time multiplied by the number of vehicles,

$$\text{total travel time} = 4.647(3500)$$
$$= 16{,}264.5 \text{ veh-min}$$

During reconstruction the performance function of route 1 is unchanged, but the performance function of route 2 is altered because of the reduction in capacity to

$$t_2 = 4 + \frac{2}{2}x_2 = 4 + x_2$$

If we assume that both routes are used, $t_1 = t_2$. Also, it is known that the total travel time is

$$t_1(q) = t_2(q)$$
$$= 16{,}264.5 \text{ veh-min}$$

Using the performance function of route 2, we obtain

$$(4 + x_2)(q) = 16.2645 \text{ (using thousands of vehicles)}$$
$$q = \frac{16.2645}{4 + x_2}$$

From $t_1 = t_2$, and $x_1 = q - x_2$ (flow conservation),

$$2 + 1.2x_1 = 4 + x_2$$
$$2 + 1.2(q - x_2) = 4 + x_2$$
$$q = 1.67 + 1.83x_2$$

Equating the two expressions for q gives

$$1.67 + 1.83x_2 = \frac{16.2645}{4 + x_2}$$

$$1.83x_2^2 + 8.99x_2 - 9.5845 = 0$$

which gives $x_2 = 0.901$, $q = 1.67 + 1.83(0.901) = 3.319$ and $x_1 = 3.319 - 0.901 = 2.418$. Since flow exists on both routes, the earlier assumption that both routes would be used is valid, and a 181(3500 − 3319) vehicle reduction in peak-hour flow is needed to ensure equality of total travel times.

Example 7.11

Two highways serve a busy corridor with a traffic demand that is fixed at 6000 vehicles during the peak hour. The service functions for the two routes are $t_1 = 4 + 5(x_1/c_1)$ and $t_2 = 3 + 7(x_2/c_2)$, where the t's are travel times in minutes, the x's are the peak-hour traffic volumes expressed in thousands, and the c's are the peak-hour route capacities expressed in thousands of vehicles per hour. Initially, the capacities of routes 1 and 2 are 4400 veh/hr and 5200 veh/hr, respectively. If a highway reconstruction project cuts the capacity of route 2 to 2200 veh/hr, how many additional vehicle hours will be added in the corridor assuming that user equilibrium conditions hold?

Solution

To determine the initial number of vehicle hours, first check to see if both routes are used,

$$t_1(6) = 10.82 \text{ min}; \quad t_2(0) = 3 \text{ min}$$

$$t_1(0) = 4 \text{ min}; \quad t_2(6) = 11.08 \text{ min}$$

Both routes are used, since $t_2(6) > t_1(0)$ and $t_1(6) > t_2(0)$. At user equilibrium, $t_1 = t_2$, so that substituting performance functions gives

$$4 + \frac{5}{4.4}(x_1) = 3 + \frac{7}{5.2}(x_2)$$

With flow conservation, $x_2 = 6 - x_1$, so that

$$4 + 1.136(x_1) = 3 + 1.346(6 - x_1)$$
$$x_1 = 2.85$$

and

$$x_2 = 6 - 2.85$$
$$= 3.15$$

The total travel time in hours is $[t_1 x_1 + t_2 x_2]/60$ or by substituting,

$$\frac{[(4 + 1.136(2.85))2850 + (3 + 1.346(3.15))3150]}{60} = 723.88 \text{ veh-hr}$$

For the reduced capacity case, the route usage check is

$$t_1(6) = 10.82 \text{ min}; \ t_2(0) = 3 \text{ min}$$
$$t_1(0) = 4 \text{ min}; \ t_2(6) = 22.09 \text{ min}$$

Again both routes are used $[t_2(6) > t_1(0)$ and $t_1(6) > t_2(0)]$. Equating performance functions (since travel times are equal) and using flow conservation $x_2 = 6 - x_1$,

$$4 + \frac{5}{4.4}(x_1) = 3 + \frac{7}{2.2}(x_2)$$
$$4 + 1.136x_1 = 3 + 3.182(6 - x_1)$$
$$x_1 = 4.19$$

and

$$x_2 = 6 - 4.19 = 1.81$$

which gives a total travel time of $([t_1 x_1 + t_2 x_2]/60)$,

$$\frac{[(4 + 1.136(4.19))4190 + (3 + 3.182(1.81))1810]}{60} = 875.97 \text{ veh-hr}$$

Thus the reduced capacity adds an additional 152.09 veh-hr (875.97 − 723.88).

7.6.3 Mathematical Programming Approach to User Equilibrium

Equating travel time on all used routes is a straightforward approach to user equilibrium, but one that can become cumbersome when many alternate routes are involved. The approach used to resolve this computational obstacle is to formulate the user equilibrium problem as a mathematical program. Specifically, user equilibrium route flows can be obtained by minimizing the following function [Sheffi 1985]

$$\min \; y(x) = \sum_n \int_0^{x_n} t_n(w) \; dw \tag{7.6}$$

where n denotes a specific route and $t_n(w)$ is the performance function corresponding to route n (w denotes flow, x_n's). This function is subject to the constraints that the flow on all routes is greater than or equal to zero ($x_n \geq 0$) and that flow conservation holds (i.e., the flow on all routes between an origin and destination sums to the total number of vehicles, q, traveling between the origin and destination, $q = \sum_n x_n$).

Formulating the user equilibrium problem as a mathematical program permits an equilibrium solution to very complex highway networks (i.e., many origins and destinations) to be readily undertaken by computer. The reader is referred to Abu-Esheh and Mannering [1986] for an application of user equilibrium principles to such a network.

Example 7.12

Solve Example 7.9 by formulating user equilibrium as a mathematical program.

Solution

From Example 7.9, the performance functions are

$$t_1 = 6 + 4x_1$$
$$t_2 = 4 + x_2^2$$

Substituting these into Eq. 7.6 gives

$$\min \; y(x) = \int_0^{x_1}(6 + 4w) \; dw + \int_0^{x_2}(4 + w^2) \; dw$$

The problem can be viewed in terms of x_2 only by noting that flow conservation implies $x_1 = 4.5 - x_2$. Substituting, we obtain

$$y(x) = \int_0^{4.5-x_2}(6 + 4w) \; dw + \int_0^{x_2}(4 + w^2) \; dw$$

$$= 6w + 2w^2 \Big|_0^{4.5-x_2} + 4w + \frac{w^3}{3} \Big|_0^{x_2}$$

$$= 27 - 6x_2 + 40.5 - 18x_2 + 2x_2^2 + 4x_2 + \frac{x_2^3}{3}$$

For arriving at a minimum, the first derivative is set to zero giving

$$\frac{dy(x)}{dx_2} = x_2^2 + 4x_2 - 20$$

$$= 0$$

which, solving for x_2, gives $x_2 = 2899$ veh/hr, the same value as was found in Example 7.9 (it can readily be shown that all other flows and travel times will also be the same as those computed in Example 7.9).

7.6.4 Theory of System Optimal Route Choice

From an idealistic point of view, one can visualize a single route choice strategy that results in the lowest possible number of total vehicle hours of travel for some specified origin/destination traffic flow. Such a strategy is known as a system optimal route choice regime and is based on the choice rule that travelers will behave such that total system travel time will be minimized even throgh travelers may be able to decrease their own individual travel times by unilaterally changing routes. From this definition it is clear that system optimal flows are not stable, since there will always be a temptation for travelers to switch to nonsystem optimal routes in order to improve their own travel times. Thus system optimal flows are generally not a realistic representation of actual traffic. Nevertheless, system optimal flows often provide for useful comparisons with the more realistic user equilibrium traffic forecasts.

The system optimal route choice rule is operationalized by the following mathematical program,

$$\min y(x) = \sum_n x_n t_n(x_n) \tag{7.7}$$

Again, this program is subject to the constraint of flow conservation ($q = \sum_n x_n$) and nonnegativity ($x_n \geq 0$).

Example 7.13

Determine the system optimal travel time for the situation described in Example 7.9.

Solution

Using Eq. 7.7 and substituting the performance functions for routes 1 and 2, we obtain

$$y(x) = x_1(6 + 4x_1) + x_2(4 + x_2^2)$$

$$= 6x_1 + 4x_1^2 + 4x_2 + x_2^3$$

From flow conservation, $x_1 = 4.5 - x_2$; therefore,

$$y(x) = 6(4.5 - x_2) + 4(4.5 - x_2)^2 + 4x_2 + x_2^3$$

$$= x_2^3 + 4x_2^2 - 38x_2 + 108$$

To find the minimum, the first derivative is set to zero,

$$\frac{dy(x)}{dx_2} = 3x_2^2 + 8x_2 - 38 = 0$$

which gives $x_2 = 2.467$ and $x_1 = 4.5 - 2.467 = 2.033$. For system optimal travel times,

$$t_1 = 6 + 4(2.033)$$

$$= 14.13 \text{ min}$$

$$t_2 = 4 + (2.467)^2$$

$$= 10.08 \text{ min}$$

which are not user equilibrium travel times, since t_1 is not equal to t_2. In Example 7.9, the total user equilibrium travel time is computed as 930 veh-hr (4500(12.4)/60). For the system optimal total travel time ($[x_1t_1 + x_2t_2]/60$),

$$\frac{[2033(14.13) + 2467(10.08)]}{60} = 893.2 \text{ veh-hr}$$

Therefore, the system optimal solution results in a systemwide saving of 36.8 veh-hr.

Example 7.14

Two roads begin at a gate entrance to a park and both take different scenic routes to a single main attraction in the park. The park manager knows that 4000 vehicles arrive during the peak hour, and he distributes these vehicles among the two routes so that an equal number of vehicles take each route. The performance functions for the routes are $t_1 = 10 + x_1$ and $t_2 = 5 + 3x_2$ with the x's expressed in thousands of vehicles per hour and the t's in minutes. How many vehicle-hours would have been saved had the park manager distributed the vehicular traffic so as to achieve a system optimal solution?

Solution

For the number of vehicle hours assuming an equal distribution of traffic among the two routes,

$$\text{Route 1: } \frac{x_1 t_1}{60} = \frac{2000[10 + (2)]}{60}$$

$$= 400 \text{ veh-hr}$$

$$\text{Route 2: } \frac{x_2 t_2}{60} = \frac{2000[5 + 3(2)]}{60}$$

$$= 366.67 \text{ veh-hr}$$

for a total of 766.67 veh-hr. With the system optimal traffic distribution, the performance functions are substituted into Eq. 7.7 giving

$$y(x) = (10 + x_1)x_1 + (5 + 3x_2)x_2$$

With flow conservation, $x_1 = 4.0 - x_2$ so that

$$y(x) = 4x_2^2 - 13x_2 + 56$$

Setting the first derivative to zero, we find that

$$\frac{dy(x)}{dx_2} = 8x_2 - 13$$

$$= 0$$

which gives $x_2 = 1.625$ and $x_1 = 4 - 1.625 = 2.375$. The total travel times are

$$\text{Route 1: } \frac{x_1 t_1}{60} = \frac{[2375(10 + 2.375)]}{60}$$

$$= 489.84 \text{ veh-hr}$$

$$\text{Route 2: } \frac{x_2 t_2}{60} = \frac{[1625(5 + 3(1.625))]}{60}$$

$$= 267.45 \text{ veh-hr}$$

which gives a total system travel time of 757.27 veh-hr or a saving of 9.38 veh-hr (766.67 − 757.29) over the equal distribution of traffic to the two routes.

7.7 PRACTICAL ISSUES IN TRAFFIC FORECASTING

As mentioned earlier, the objective of this chapter was to convey the basic principles underlying traffic forecasting. One can readily imagine that the actual application of the various methods presented in this chapter to a real-world traffic forecasting problem requires that a number of important practical issues be addressed. Such issues include: (1) the geographical definition of origin and destination areas, (2) the definition of a household as a decision-making unit, (3) the possibility of trip chaining (i.e., combining trips such as stopping to shop on the way to work), (4) the large number of alternate routes available between most origins and destinations, (5) specifying appropriate route performance functions, (6) the equilibrium problem in nonroute choice models (e.g., travel time effects on mode/destination and trip generation choices), and (7) the long-term effect of traveler decisions on residential and commercial locations (see Fig. 7.1). These issues add still other dimensions to an already difficult problem. The reader is referred to other references [Meyer and Miller 1984; Morlok 1978] for descriptions of how traffic forecasting efforts have historically attempted to address these important practical issues.

APPENDIX 7A

METHOD OF LEAST SQUARES REGRESSION

Least squares regression is a popular method of developing mathematical relationships from empirical data. As mentioned in the text, it is a method that is well suited to the estimation of trip generation models. To illustrate the least squares approach, consider the hypothetical trip generation data presented in Table 7A.1, which could have been gathered from a typical survey of travelers.

To begin formalizing a mathematical expression, note that the objective is to predict the number of shopping trips made on a Saturday for each household, i; this number is referred to as the dependent variable (Y_i). This prediction is to be a function of the number of people in household i (z_i), which is referred to as the independent variable. A simple linear relationship between Y_i and z_i is

$$Y_i = b_0 + b_1 z_i \tag{7A.1}$$

where the b's are coefficients to be determined and Y_i is the number of shopping trips predicted by the equation. Ideally, we want to determine the b's in Eq. 7A.1 that will give predictions of the number of shopping trips (Y_i's) that are as close as possible to the actual observed number of shopping trips (Y_i's, as shown in Table 7A.1). The difference or deviation between the observed and predicted number of shopping trips can be expressed mathematically as

$$\text{deviation} = Y_i - (b_0 + b_1 z_i) \tag{7A.2}$$

TABLE 7A.1
Example of Shopping Trip Generation Data

Household Number i	Number of Shopping Trips Made All Day Saturday Y_i	People in Household i z_i
1	3	4
2	1	2
3	1	3
4	5	4
5	3	2
6	2	4
7	6	8
8	4	6
9	5	6
10	2	2

Graphically, such deviations are illustrated in Fig. 7A.1 for two groups of b_0 and b_1 values. In the first illustration of this figure, $b_0 = 1.5$ and $b_1 = 0$, which implies that the number of household members does not affect the number of shopping trips made. The second illustration has $b_0 = 1.5$ and $b_1 = 0.5$ and, as can readily be seen, the deviations (differences between the points representing the observed number of shopping trips and the line representing the equation, $b_0 + b_1 z_i$) are reduced relative to the first illustration. These two illustrations suggest the need for some method of determining the values of b_0 and b_1 that

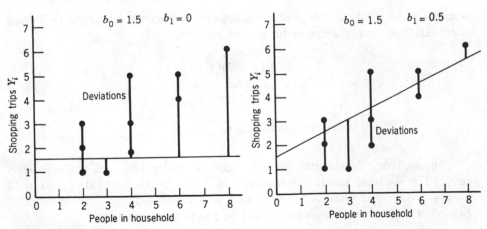

FIGURE 7A.1
Illustration of deviations.

produce the smallest possible deviations relative to observed data. Such a method can be solved by a mathematical program whose objective is to minimize the sum of the square of deviations, or

$$\min \, y(b_0, b_1) = \sum_i (Y_i - b_0 - b_1 z_i)^2 \tag{7A.3}$$

The minimization is accomplished by setting partial derivatives equal to zero,

$$\frac{\partial y}{\partial b_0} = -2\sum_i (Y_i - b_0 - b_1 z_i)$$

$$= 0$$

$$\frac{\partial y}{\partial b_1} = -2\sum_i z_i (Y_i - b_0 - b_1 z_i)$$

$$= 0$$

giving

$$\sum_i (Y_i - b_0 - b_1 z_i) = 0$$

$$\sum_i z_i (Y_i - b_0 - b_1 z_i) = 0$$

or

$$\sum_i Y_i - nb_0 - b_1 \sum_i z_i = 0$$

$$\sum_i z_i Y_i - b_0 \sum_i z_i - b_1 \sum_i z_i^2 = 0$$

where n is the total number of observations (or, in this case, households). Solving these equations simultaneously for b_0 and b_1 gives

$$b_1 = \frac{\sum_i (z_i - \bar{z})(Y_i - \bar{Y})}{\sum_i (z_i - \bar{z})^2} \tag{7A.4}$$

$$b_0 = \bar{Y} - b_1 \bar{z} \tag{7A.5}$$

This approach to determining the values of estimable coefficients (b's) is referred to as least squares regression, and it can be shown that for the data values given in Table 7A.1, the lowest deviations between the number of predicted and actual shopping trips will be given by the equation,

$$Y_i = 0.33 + 0.7 z_i$$

When many coefficient values (b's) must be determined, a matrix representation of the least squares solution is appropriate,

$$\mathbf{B} = (\mathbf{z'z})^{-1}\mathbf{z'Y} \tag{7A.6}$$

where $\mathbf{B}$ is a vector of b_n's, $\mathbf{z}$ is a matrix of characteristics, and $\mathbf{Y}$ is the vector of dependent variables. For additional information on least squares regression, the reader is referred to Neter and Wasserman [1973].

NOMENCLATURE
FOR
CHAPTER 7

b_n estimated coefficients

P probability of an alternative being selected

q total origin to destination traffic flow

s notation for the set of available alternatives

T_{ij} number of household trips generated per unit time

t_a travel time on route a

U specifiable portion of an alternative's utility

V total alternative utility

w route flow operative for x_a

x_a traffic flow on route a

Y_i dependent variable

$y(\cdot)$ mathematical objective function

z household or alternative characteristic

ε unspecifiable portion of an alternative's utility (assumed to be a random variable)

REFERENCES

1. M. Meyer and E. Miller, *Urban Transportation Planning: A Decision-Oriented Approach*, McGraw-Hill Book Company, New York, 1984.

2. C. S. Papacostas, *Fundamentals of Transportation Engineering*, Prentice-Hall, Inc., Englewood Cliffs, N.J., 1987.

3. L. Haefner, *Introduction to Transportation Systems*, Holt, Rinehart and Winston, New York, 1986.

4. E. Morlok, *Introduction to Transportation Engineering and Planning*, McGraw-Hill Book Company, New York, 1978.

5. Y. Sheffi, *Urban Transportation Networks: Equilibrium Analysis with Mathematical Programming Models*, Prentice-Hall, Inc., Englewood Cliffs, N.J., 1985.

6. M. Ben-Akiva and S. Lerman, *Discrete Choice Analysis: Theory and Application to Travel Demand*, MIT Press, Cambridge, Mass., 1985.

7. H. Mahmassani and G.-L. Chang, "On Boundedly Rational User Equilibrium in Transportation Systems," *Transportation Science*, Vol. 21, No. 2, May 1987.

8. S. Abu-Eisheh and F. Mannering, "A Discrete/Continuous Analysis of Commuters' Route and Departure Time Choice," *Transportation Research Record*, 1138, 1988.

9. F. Mannering, "Poisson Analysis of Commuter Flexibility in Changing Routes and Departure Times," *Transportation Research*, Vol. 22B, No. 2, 1989.

10. D. McFadden, "Econometric Models of Probabilistic Choice," in *Structural Analysis of Discrete Data with Econometric Applications*, Manski and McFadden, eds., MIT Press, Cambridge, Mass., 1981.

11. J. Wardrop, "Some Theoretical Aspects of Road Traffic Research," *Proceedings, Institution of Civil Engineers II*, Vol. 1, 1952.

12. S. Abu-Eisheh and F. Mannering, "Traffic Forecasting for Small- to Medium-Sized Urban Areas," *ITE Journal*, Vol. 56, No. 10, October 1986.

13. J. Neter and W. Wasserman, *Applied Linear Statistical Models*, Richard Irwin, Inc., Homewood, Ill., 1973.

PROBLEMS

7.1. A large retirement village has a total retail employment of 100. All 1700 of the households residing in this village consist of two nonworking family members with household incomes of $20,000. Assuming that shopping and social/recreational trip rates both peak during the same hour (for exposition purposes) predict the total number of peak-hour trips generated by this village using the trip generation models of Examples 7.1 and 7.2.

7.2. Consider the retirement village described in Problem 7.1. Determine the amount of additional retail employment (in the village) necessary to reduce the total predicted number of peak-hour shopping trips to 100.

7.3. A large residential area has 1500 households with an average household income of $15,000, an average household size of 5.2, and, on the average, 1.2 working members. Using the model described in Example 7.2 (assuming that it was estimated using zonal averages instead of individual households), predict the change in the number of peak-hour social/recreational trips if employment in the area increased by 20 percent and household income by 10 percent.

7.4. If small express buses leave the origin described in Example 7.4 and all are filled to their capacity of 15 travelers, how many work trip vehicles leave from origin to destination in Example 7.4 during the peak hour?

7.5. Consider the conditions described in Example 7.4. If an energy crisis doubles the cost of the auto modes (i.e., drive alone and shared rides) and bus costs are not affected, how many workers will take each mode?

7.6. It is known that 4000 automobile trips are generated in a large residential area from noon to 1:00 p.m. on Saturdays for shopping purposes. Four major shopping centers have the following characteristics.

Shopping Center Number	Distance from Residential Area (miles)	Commercial Floor Space (in thousands of ft^2)
1	4	200
2	9	150
3	8	300
4	14	600

If a logit model is estimated with coefficients -0.283 for distance and 0.0172 for commercial space (in thousands of ft^2), how many shopping trips will be made to each of the four shopping centers?

7.7. Consider the shopping trip situation described in Problem 7.6. Suppose that shopping center 3 goes out of business and shopping center 2 is expanded to 500,000 square feet of commercial space. What would be the new distribution of the 4000 Saturday afternoon shopping trips?

7.8. If shopping center 3 is closed (see Problem 7.7), how much commercial floor space is needed in shopping centers 1 and 2 to ensure that each of them have the same probability of being selected as shopping center 4?

7.9. Consider the situation described in Example 7.6. If the construction of a new freeway lowers auto and transit travel times to shopping center 2 by

20 percent, determine the new distribution of shopping trips by destination and mode.

7.10. Consider the conditions described in Example 7.6. Heavily congested highways have caused travel times to increase to shopping center 2 by 4 min for both auto and transit modes (travel times to shopping center 1 are not affected). In order to attract as many total trips (auto and transit) as it did before the congestion, how much commerical floor space must be added to shopping center 2 (given that the total number of departing shopping trips remains at 900)?

7.11. A total of 700 auto-mode social/recreational trips are made from an origin (residential area) during the peak hour. A logit model estimation is made and three factors were found to influence the destination choice: (1) population at the destination in thousands (coefficient = 0.2), (2) distance from origin to destination in miles (coefficient = −0.15), and (3) square feet of amusement floor space (e.g., movie theaters, video game centers, etc.) in thousands (coefficient = 0.09). Four possible destinations have the following characteristics:

	Population (in thousands)	Distance from Origin (in miles)	Amusement Space (in thousands of ft^2)
Destination 1	15.5	12	5
Destination 2	6.0	8	10
Destination 3	0.8	3	8
Destination 4	5.0	11	15

Determine the distribution of trips among possible destinations.

7.12. Consider the situation described in Problem 7.11. If a new 15,000 square foot arcade center is built at destination 3, determine the distribution of the 700 peak-hour social/recreational trips.

7.13. Note that with the situation described in Example 7.7, 26.6 percent, (110 + 23)/500, of all social/recreational trips are destined for destination 1. If the total number of trips remains constant, how much additional amusement floor space would have to be added to destination 1 to have it capture 40.0 percent of the total social/recreational trips?

7.14. Consider the situation described in Problem 7.11. Destination 2 currently attracts 146 of the 700 social/recreational trips. If the total number of trips remains constant, determine the amount of amusement floor space that must be added to destination 2 to attract a total of 250 social/recreational trips.

7.15. Two routes connect an origin and a destination and the flow is 15,000 veh/hr. Route 1 has a performance function $t_1 = 4 + 3x_1$, and route 2 has a function of $t_2 = b + 6x_2$, with the x's expressed in thousands of vehicles per hour and the t's in minutes.

    (a) If the user equilibrium flow on route 1 is 9780 veh/hr, determine the free-flow speed on route 2 (i.e., b) and equilibrium travel times.

    (b) If population declines reduce the number of travelers at the origin and the total origin–destination flow is reduced to 7000 veh/hr, determine user equilibrium travel times and flows.

7.16. An origin-destination pair is connected by a route with a performance function $t_1 = 8 + x_1$, and another with a function $t_2 = 1 + 2x_2$ (x's in thousands of vehicles per hour and t's in minutes). If the total origin-destination flow is 4000 veh/hr, determine user equilibrium and system optimal route travel times, total travel time (in vehicle-minutes) and route flows.

7.17. Because of the great increase in vehicle-hours caused by the reconstruction project in Example 7.11, the state transportation department decides to regulate the flow of traffic on the two routes (until reconstruction is complete) to achieve a system optimal solution. How many vehicle-hours will be saved during each peak-hour period if this strategy is implemented and travelers are not permitted to achieve a user equilibrium solution?

7.18. For Example 7.10, what reduction in peak-hour traffic demand is needed to ensure an equality of total vehicular travel time (in vehicle-minutes) assuming a system optimal solution before and during reconstruction?

7.19. Two routes connect an origin and a destination. Routes 1 and 2 have performance functions $t_1 = 2 + x_1$ and $t_2 = 1 + x_2$, where the t's are in minutes and the x's are in thousands of vehicles per hour. The travel time on the routes is known to be in user equilibrium. If an observation on route 1 finds that the gaps between 40 percent of the vehicles is less than 5 sec, estimate the volume and average travel times on the two routes. (*Hint:* Assume a Poisson distribution of vehicle arrivals as discussed in Chapter 5.)

7.20. Three routes connect an origin and a destination with performance functions $t_1 = 8 + 0.5x_1$, $t_2 = 1 + 2x_2$, $t_3 = 3 + 0.75x_3$, with the x's expressed in thousands of vehicles per hour and the t's expressed in minutes. If the peak-hour traffic demand is 3000 vehicles, determine user equilibrium traffic flows.

7.21. Two routes connect a suburban area and a city with route travel times (in minutes) given by the expressions $t_1 = 6 + 8(x_1/c_1)$ and $t_2 = 10 + 3(x_2/c_2)$, where the x's are expressed in thousands of vehicles per hour and the c's are the route capacities in thousands of vehicles per hour.

Initially, the capacities of routes 1 and 2 are 4000 and 2000 veh/hr, respectively. A reconstruction project on route 1 reduces the capacity to 3000 veh/hr, but total traffic demand is unaffected. Observational studies note a 35.28 sec increase in average travel time on route 1 and a 68.5 percent increase in flow on route 2 after reconstruction begins. User equilibrium conditions exist before and during reconstruction. If both routes are always used, determine equilibrium flows and travel times before and after reconstruction begins.

7.22. Three routes connect an origin and destination with performance functions $t_1 = 2 + 0.5x_1$, $t_2 = 1 + x_2$, $t_3 = 4 + 0.2x_3$ (t's in minutes and x's in thousands of vehicles per hour). Determine user equilibrium flows if the total origin to destination demand is (a) 10,000 veh/hr and (b) 5000 veh/hr.

7.23. For the routes described in Problem 7.22, what is the minimum orgin to destination traffic demand (in vehicles per hour) that will ensure that all routes are used (assuming user equilibrium conditions).

7.24. An urban freeway has five lanes, four of which are unrestricted (open to all vehicular traffic) and one restricted lane that can be used only by vehicles with two or more occupants. The performance functions, between origin and destination, are $t_r = 4 + 2x_r$ for the restricted lane and $t_u = 4 + 0.5x_u$ for the unrestricted lanes, combined (t's in minutes and x's in thousands of vehicles per hour). During the peak hour, 2000 vehciles with one occupant and 2000 vehicles with two occupants depart for the destination. Determine the distribution of traffic among lanes such that total person-hours is minimized and compare the savings in person-hours relative to a user equilibrium distribution of traffic among lanes.

Index